Introduction to Small-scale Geological Structures

Introduction to Small-scale Geological Structures

Gilbert Wilson

*Department of Geology,
Imperial College of Science and Technology*

in collaboration with

J. W. Cosgrove

London
GEORGE ALLEN & UNWIN
Boston   Sydney

© G. Wilson, 1982
This book is copyright under the Berne Convention. No reproduction without permission. All rights reserved.

**George Allen & Unwin (Publishers) Ltd,
40 Museum Street, London WC1A 1LU, UK**

George Allen & Unwin (Publishers) Ltd,
Park Lane, Hemel Hempstead, Herts HP2 4TE, UK

Allen & Unwin Inc.,
9 Winchester Terrace, Winchester, Mass 01890, USA

George Allen & Unwin Australia Pty Ltd,
8 Napier Street, North Sydney, NSW 2060, Australia

First published in 1982

British Library Cataloguing in Publication Data

Wilson, Gilbert
  Introduction to small-scale geological structures
1. Geology, Structure
I. Title
551.8   QE601
ISBN 0-04-551051-2
ISBN 0-04-551052-0 Pbk

Library of Congress Cataloging in Publication Data

Wilson, Gilbert
  Introduction to small-scale geological structures
  Bibliography: p.
  Includes index.
  1. Geology, Structural.   I. Cosgrove, J. W. (John W.)   II. Title.
  QE601.2.W54   551.8   81-14954
  ISBN 0-04-551051-2   AACR2
  ISBN 0-04-551052-0 (pbk.)

Set in 10 on 12 point Times by Typesetters (Birmingham) Limited,
and printed in Great Britain
by Hazell Watson and Viney Limited, Aylesbury, Bucks

Preface

The small-scale structures referred to in this publication are those structures of tectonic origin that can be observed with the naked eye in the field. Their scale varies broadly between that of the hand-specimen to that of the exposure, or even mountainside. Such structures are the visible effects of rock deformation caused by local stresses and movements which have been induced in the rocks by external tectonic forces of possibly unknown origin. Recognition of these minor structures, and appreciation of their origin and significance assist the field geologist to elucidate the larger-scale geological structures of his area. Commonly some can be used in deciphering the order of stratigraphic succession in regions of strongly-folded unfossiliferous beds; and, in ground which has suffered superposed tectonic movements, the minor structures may provide evidence of successive phases or events in the tectonic history. The work contains descriptions of the more common varieties of small-scale tectonic structures, the different ways in which these structures may have been formed, and the limitations of the conclusions which can be drawn from their observation in the field.

Gilbert Wilson
June 1981

Acknowledgements

An outline of much of the material given in this book was delivered at the 'Cinquieme Conference Gustave Dewalque' to the Société Géologique de Belgique in 1958 and was published in the annals of the society in 1961. The foundations of much that I have written were laid in the lectures on structural geology by Professors C. K. Leith and W. J. Mead at the University of Wisconsin many years ago. The principles which they expound run through the whole of this work, in places disguised in modern jargon, elsewhere modified by more recent advances in knowledge, but it was they who instilled in me the importance of minor structures in the elucidation of the major structures in the field.

In preparing the 1961 version of the work I had the aid of many of my colleagues in the Department of Geology, Imperial College, London. The late Professor H. H. Read, FRS, Professor John Sutton, FRS and his wife, Professor Janet Watson, FRS read much of the script in its various stages and made numerous helpful suggestions and the advice of Professor J. G. Ramsay, FRS with his detailed understanding of the effects of superposed tectonics in complex metamorphosed terrains was of the greatest assistance to me.

At the beginning of 1980 a suggestion was made that someone with a more modern outlook in structural geology be asked in to assist in the authorship of this work and I was delighted to obtain from Dr John Cosgrove of the Department of Geology, Imperial College permission to include his name in this publication. His help came at a moment of great difficulty in my life when I had suffered from a stroke and was unable to concentrate on more recent publications with which he was more fully familiar.

Permission to reproduce Figures 6.1b, 8.5c and 13.2c was kindly granted by the Council of the Geological Society of London. The Council of the Geologists' Association have also permitted me to reproduce Figures 2.1, 4.3, 5.5c, 6.6 and 11.2; the drawing, Figure 5.2, was made from a photograph by the late Dr A. J. Bull, which appeared in the *Proceedings* of the Association, vol. 33, Plate 7A. Figures 5.5a & b and Figure 5.7b, which was redrawn from Flinn's original diagram (1952), have been included through the kindness of the Editor of the Geological Magazine and Professor Derek Flinn.

Lastly, I would like to record my debt to two people: the first, my wife, but for whose assistance in the translation of my notes into French, the original lecture could never have been delivered; the second is the late Professor Paul Fourmarier, who continually helped and encouraged me in this work. To him in particular, I express my sincerest appreciation and gratitude.

Contents

		page
Preface		vii
Acknowledgements		viii
1	Introduction	1
2	Stress and strain	3
3	Structural symmetry	10
4	Structure and stratigraphical succession	17
5	Structures in brittle rocks: tension fractures and shear zones	24
6	Rock cleavage and schistosity: generalities	35
7	Fracture cleavage and strain-slip cleavage	48
8	Flow cleavage, schistosity and lineation	61
9	Boudinage	72
10	Drag-folds and parasitic folds	80
11	Mullion and rodding structures	86
12	Superposed minor structures	94
13	Minor structures and large-scale tectonics	104
14	Conclusions	111
Bibliography		113
Index		126

1 Introduction

More than 80 years ago, Professor Charles Lapworth (1893) considered that two of the life-stages of a geological formation, detrition and deposition, had been studied in the light of present-day processes and that the third stage, deformation, should be similarly studied. Though deformation of the rocks cannot be studied 'in the light of present-day processes' in the field, much has been learned by experiment in the laboratory and by the application of mechanical principles to the interpretation of the structures observed. Since Lapworth wrote the words above, great advances have been made in the 'study of deformation', that is in **structural geology** and **tectonics**. Such advances – for the most part made by specialists who are inclined towards mathematical and mechanical studies – are commonly presented to the reader in a form which frightens the non-specialist to such an extent that he either leaves the subject entirely alone, or he loses himself in a maze of detail and jargon.

The object of this work is to try to present, to the ordinary field geologist, the importance and significance of the many small-scale structures which can be seen in the rocks. Such small-scale structures, to my mind, are those which can readily be seen in the field, but are too small to record (other than diagrammatically) on a map of scale 1 : 10 000 or less. As Professor Eugène Wegmann would say, they are of the scale of the exposure or of the hand-specimen. The significance of these structures is four-fold: they may be used in the elucidation of the geometry of the larger-scale structures and the stratigraphical succession of the beds; they may be used as 'tectonic weather-cocks' for indicating the directions and sense of the local movements which have affected the rocks; they may give an indication of the distribution of the stresses which were responsible for the deformation, and they may indicate a time sequence of different phases of deformation in which case the structures assist in deciphering the tectonic history.

It may be noticed that, in the foregoing paragraphs, I have not used the terms structural geology and tectonics as if they were synonymous. To me, structural geology is primarily concerned with the geometry of the rocks; and tectonics, according to a definition by Professor H. H. Read, is '. . . the structure considered in relation to the forces and movements that have operated in a region' (Read 1949a). One is the photograph, the other the cinematograph; and even though Goguel

(1948) tells us that 'La cinématique n'est que la géometrie dans le temps', we are, by the introduction of *time*, making the rocks live. This means that the study of the structures can be changed from a problem in geometry or dynamics into one of historical geology.

Once the rocks have left their original environment of sedimentation, where the conditions under which they were born are indicated by their facies, their subsequent history is written at first in their diagenesis, and then in their deformation and metamorphism. The aim of structural geology and tectonics is not just the study of the mechanisms of rock-deformation; rather it is to try and continue the geological history, shown by the rocks of an area, beyond the period when the stratigrapher and sedimentologist have had to stop. Before this can be done, however, an understanding of the modes of formation of the various rock-structures is essential to the worker who proposes to interpret what he observes: just as the principles of evolution are important to the palaeontologist. In the next chapter, therefore, I propose to outline briefly the relationship between the application of forces, and the deformations which they would cause, together with simple examples of the resulting structures. Later I shall describe and discuss the more common varieties of geological structures and shall endeavour to elucidate their tectonic significance.

2 Stress and strain

Let us, to begin with, consider a spherical rock element in a static region of the Earth's crust. The stress (force/unit area) acting on the element will be the same in all directions. Such a stress condition is termed **hydrostatic** and has been defined by Anderson (1951, p. 13) as the **standard state**. However, the stress state in the Earth's crust is often not truly hydrostatic; nevertheless any balanced system of stresses, whether caused by compressional, tensional or torsional forces, can be resolved into three principal stresses at right angles to each other (Hills 1953 p. 23 et seq., Price 1966). These three principal stresses can be symbolised as $P_{maximum}$, $P_{intermediate}$ and $P_{minimum}$. Alternatively they are often referred to by using the Greek letters σ_1, σ_2 and σ_3 respectively.

In general, P_{max} and P_{min} are both compressive and it is doubtful whether truly tensional stresses are really active in the Earth's crust. However, this does not mean that rocks cannot fail by tension. The factor which determines whether a rock will fail by tension or by shear failure is the **stress-difference** or **differential stress** ($\sigma_1-\sigma_3$) (Price 1975). High differential stress favours shear failure, low differential stress favours tensile failure.

The stress state at any instance can be represented by these three mutually perpendicular principal stresses, and the ellipsoid, whose axes are proportional in length and parallel to the principal stress vectors, is known as the **stress ellipsoid**. The stress ellipsoid corresponding to the standard state of hydrostatic stress is a sphere. The effect of a stress state with three unequal principal stresses (i.e. a non-hydrostatic stress state) on a rock is best exemplified by considering geometrically the change in form of an originally spherical element. This sphere, when subjected to a non-hydrostatic stress would deform into a triaxial ellipsoid (Fig. 2.1). This ellipsoid is referred to as the **strain ellipsoid** or **ellipsoid of deformation**.

For incremental deformation, i.e. the deformation with which engineers are generally concerned, and for certain finite deformations, e.g. 'pure shear', the three axes of the strain ellipsoid AA, BB and CC, correspond respectively in direction but not in lengths to the three principal stresses, P_{min}, P_{int} and P_{max}; CC and AA are the directions of minimum and maximum extensions respectively (Fig. 2.2a). In other types of deformation, e.g. 'simple shear', the stress ellipsoid and finite strain ellipsoid are non-coaxial (Fig. 2.2b). Simple shear deformation

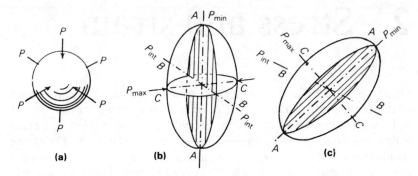

Figure 2.1 A spherical element under hydrostatic pressure (a), being deformed into a triaxial ellipsoid, (b) and (c), by variations in the magnitudes and orientations of the principal stresses (after Wilson 1946).

may develop when a torsional force, or couple, is primarily responsible for the stress induced in the rock. Such a couple of forces develops quite commonly in nature, and it often results during the folding of strata from the frictional force which resists slip of the beds along planes of stratification. It may also be transmitted through the rocks in the neighbourhood of a zone of faulting or thrusting, or beneath a *traîneau écraseur* or crushing sledge, formed by an important and ponderous mass of rock being driven over and crushing an underlying fold or other structure beneath.

A couple of this style, if it acted on the top and bottom of a free element, would tend to make the latter rotate, like a roller lying between two surfaces which move in opposite directions (Fig. 2.3a). Rarely such rotation may be recognised in the rocks, for instance, where a more rigid crystal is embedded in a softer schist which is being deformed. More commonly it does not occur because in stratified rocks in particular, the element is held in place by the beds above and below it, like the crosshead of a connecting rod or a piston between its guides. This indicates that a second couple, acting at right-angles to, and in the plane of the primary couple, must have been introduced in order to prevent the rotation (Fig. 2.3b). Resolution of these two couples shows that their dynamic effect is equivalent to two principal stresses P_{max} and P_{min} acting at 45° to the shear direction (Fig. 2.3b). Initially, that is, for the first increment of deformation, the maximum and minimum axes of the strain ellipsoid will also be inclined at 45° to the direction of shear. These two axes will be parallel to P_{min} and P_{max} respectively (Figs 2.2b, 2.3b). However, with progressive deformation the finite strain ellipsoid will rotate about its intermediate axis such that the direction of maximum

Stress and strain

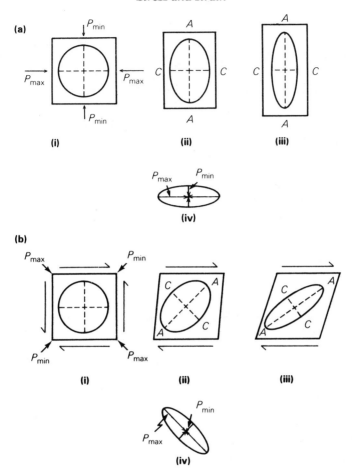

Figure 2.2 (a) (i), (ii), (iii) Stages in the deformation of an original circle and square by 'pure shear'. The original circle on the undeformed square is deformed into an ellipse, the strain ellipse, the major and minor axes of which are parallel to and proportional in length to the maximum and minimum principal extensions respectively. (iv) The stress ellipse. The maximum and minimum principal compression are parallel to the major and minor axes respectively. It can be seen that the principal axes of stress and strain always remain parallel during pure shear deformation. (b) (i), (ii), (iii) Stages in the deformation of an original circle and square by 'simple shear'. The original circle is deformed into an ellipse, the strain ellipse, the major and minor axes of which are parallel to and proportional in length to the maximum and minimum principal extensions respectively. (iv) The stress ellipse. The maximum and minimum principal compressions are parallel to the major and minor axes respectively. In simple shear the principal axes of strain and stress are only parallel for the first increment of strain, (b) (ii), after which the principal strain axes rotate away from principal stress axes, (b) (iii).

extension ($A-A$) rotates towards the direction of shear. The stress- and strain-ellipsoids will then no longer be coaxial. The effect of the rotation of the finite strain ellipsoid on the geometry of structures that formed during the development of simple shear is well illustrated by the formation of sigmoidal tension gashes *en échelon* discussed in Chapter 5 and shown in Figure 5.3. Structures formed as a result of simple shear will be inclined to the direction of shear and will have monoclinic symmetry.

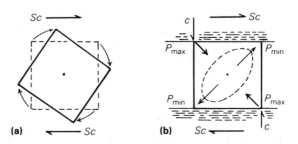

Figure 2.3 (a) Stress-couple Sc acting on a free element; (b) stress-couple Sc balanced by an induced secondary couple cc.

If the forces are great enough, the rocks on which they are acting will either deform or, if their strengths be exceeded, they will rupture. The deformation of the rock which has been subjected to such conditions depends on several factors:

(a) The difference in magnitude between the maximum and minimum principal stresses, referred to as the stress-difference, or differential stress.
(b) The strength of the rock, which in turn depends on:
 (i) Its lithology, for example, sandstone is more resistant to deformation than shale.
 (ii) The ambient or confining pressure, which is largely a factor of depth of burial. Thus, under shallow depth conditions limestone has a high rigidity, but at depth it becomes ductile and may flow.
 (iii) The presence of liquids within the rock. Fluids in the pore-spaces of the rocks commonly have a high hydrostatic pressure which may be sufficiently large to cause hydraulic fractures to develop. Pore-space fluids at a high temperature also promote recrystallisation and metamorphism, and in many cases reduce the resistance of the rock to deformation.

(c) The rate of application (strain rate) and the duration of time over which the rock has been subjected to the external forces.

It should, however, be recognised that the final deformations seen in the rocks in the field, are generally not the results of a single dynamic action alone. Further displacements during and after the initial formation of a structure may have, in fact probably have, taken place. In consequence, it must be borne in mind that the structures one now sees are the products of rock failure resulting from the initial stresses, combined with or complicated by subsequent movements.

The effect of an all-round increase in the confining pressure on a rock, equivalent to burial at depth, is to raise its strength far above that which it would have when tested under ordinary laboratory conditions. At the same time it allows the rock to deform to a considerable extent without breaking even though the stress-difference might be so great that the material would have failed by rupture under ordinary circumstances. This deformation at depth is increased, or rather facilitated, by the presence of pore fluids, either connate or magmatic, in the rock; and also by the length of time over which the stress is active. The effects of these factors have been demonstrated by numerous experiments in which rock specimens have been tested under conditions of varying pressure and for different lengths of time. The pioneer work was performed under the direction of F. D. Adams at McGill University, Montreal: Adams and Nicholson (1901); Adams (1912); Adams and Bancroft (1917); and also von Karman (1911). More carefully controlled experiments have since been made by Griggs (1936, 1938, 1939, 1940, 1942), Price (1966), Rutter (1974) and Tullis et al. (1979). An excellent discussion of the problem was given in de Sitter (1956, 1964, p. 43).

The experimental results indicate that, given time and high retaining pressures, the rocks will deform plastically, or by some manner analogous to elastico-viscous flow. Carey (1954) considered that the deformation could be expressed by the equation:

$$S = \frac{P}{\mu} + f(P) + t^{1/3}\beta + \frac{Pt}{\eta}$$

Total strain (S) = elastic + plastic + transient + viscous strain. S = shear strain; P = shear stress; t = time; μ = rigidity; η = coefficient of viscosity and β is a constant.

This formula, the geological implications of which Carey discusses at length, suggests that for stresses of short duration the rocks behave as rigid elastic solids; for stresses of longer duration they deform plastically; but over very long periods of time the last term in the equation becomes dominant over the other three, and the deformation is that of a viscous fluid, and 'creep' occurs. As de Sitter and others have pointed out, there is very strong evidence to suggest that, even when subjected to a stress over very long periods, rocks possessed a certain initial rigidity. They will not show a permanent deformation by flow, or by any other mechanism, until a certain threshold stress has been exceeded. This also agrees with Scheidegger's (1958) analysis of the 'mechanics of deformation' in the Earth's crust and of the mantle which lies below it.

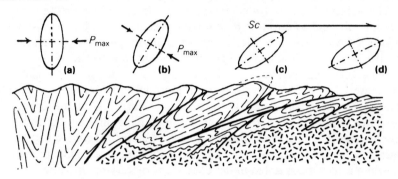

Figure 2.4 Variations in deformation in an ideal orogen.

The orientations of the principal stresses and of the structures for which they are responsible will vary from place to place depending on local conditions. For example, in the centre of an ideal orogen (Fig. 2.4), the forces acting on the rocks can be considered to be dominantly compressional, and the main direction of relief (P_{min}) will be vertical. The principal compressive stress will thus be more or less horizontal, and will act like a vice. The larger folds, in consequence, would tend to be upright and symmetrical, and the deformation as a whole can be represented by the strain ellipsoid whose shortest axis is horizontal and whose longest axis is vertical (Fig. 2.4a). Away from the centre of our orogen the direct compression is replaced by a shearing force or couple. This causes the principal stresses to be rotated in space, so that the maximum stress (P_{max}) is no longer horizontal and the direction of relief (P_{min}) is itself inclined. The deformation will conform to these changes in orientation of the system of stresses (Fig. 2.4c); and the extent to which the structures are overturned, or their asymmetry, will depend on the attitude of the principal stresses. Near the margins of our orogen the rocks can move more freely, so that in addition to the purely dynamic aspect of deformation, we must take into consideration the effect of movement. This may rotate the structures still further, so that they tend to become more and more asymmetrical (Fig. 2.4d). Here we have the apparently anomalous condition in which P_{max} appears to have been acting in a nearly vertical direction.

In the section across the orogen, we see that the forms of the structures, the forces and the movements are closely related; and that the symmetry of the first reflects that of the other two. In the centre the symmetry is more or less orthorhombic, near the margins it becomes markedly monoclinic − the plane of the page acting as a plane of symmetry. If the front of the range, that is its third or longitudinal dimension which is not shown, were oblique to the plane of the section, or were undulating, then we would find that the structures, and the forces and movements responsible for them, were asymmetric in all three directions, and so would be triclinic.

This concept of symmetry in rock-structures mentioned in the last paragraph, and its relation to the kinematics and dynamics that were responsible for the deformation, have been largely developed by Professor Bruno Sander of Austria and his followers. The significance of this work has been summarised by Turner (1953) who considered that:

> In the science of *Gefügekunde* as developed by Sander, perhaps the most far-reaching assumption is that the symmetry of deformed rock fabric reflects the symmetry of the stress and movement involved in deformation. This assumption has received confirmation from a number of sources.

3 Structural symmetry

The symmetry of any tectonic structure can in general be described in terms analogous to those used in crystallography. Thus we may speak of a piece of homogeneous rock, e.g. an unfoliated granite or gabbro, as being **isotropic**.

A pebble, which has been uniformly flattened or extended parallel to a single axis into an oblate or prolate ellipsoid, would have **axial symmetry**. Any number of planes of symmetry could be constructed to pass through the figure, provided they contained the single axis or were normal to it.

A structure is said to have **orthorhombic symmetry** if it contains two — not necessarily three — planes of symmetry at right angles to each other. Thus, a system of parallel symmetrical folds have orthorhombic symmetry in the structural sense, though not in the true crystallographic sense. One plane of symmetry will coincide with the folds' axial planes which are parallel and the other will be at right angles to the axes themselves (Fig. 3.1a).

The two intersecting zones of tension gashes *en échelon* shown in Figure 3.1b, also form an orthorhombic structure.

Monoclinic symmetry is the most common variety of structural symmetry in tectonically deformed rocks (Fig. 3.1c). The structure has a single plane of symmetry which, in an ordinary flexural fold, is at right angles to the fold-hinge. In a rock with slaty cleavage, it would be at right angles to the line of intersection of bedding and cleavage (Fig. 3.1d).

Triclinic structures have no planes of symmetry, for example the 'corkscrew-shaped folds' mentioned by de Margerie and Heim (1888, p. 118). These nonsymmetrical structures may arise from the oblique intersection of two sets of structures of different ages, each of which may originally have had a more simple form of symmetry (Fig. 3.1e). Alternatively, they might be formed by the uneven development or twisting of a structure during a single movement.

Structural symmetry commonly varies in different portions of a big structure. For example, parasitic or small monoclinic folds may occur on the limbs of a major orthorhombic fold; orthorhombic zig-zag folds may form the hinge zone of a large monoclinic fold; or a stretched belemnite with axial symmetry may lie in a zone of thrusting which, when viewed as a whole, is monoclinic. The symmetry shown by minor structures in isolated exposures does not always reflect the symmetry of the major

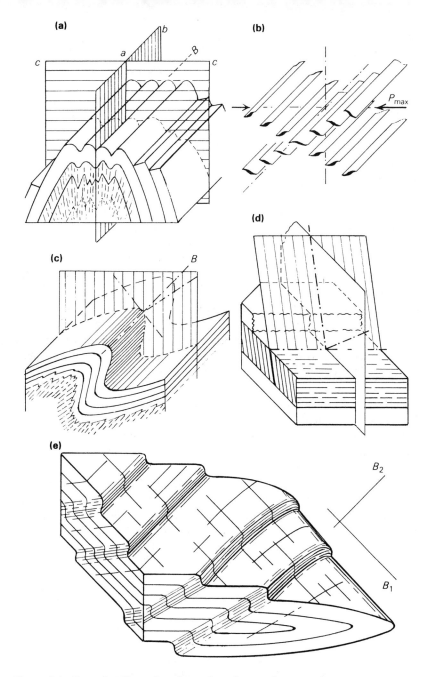

Figure 3.1 Examples illustrating the various forms of structural symmetry. (a), (b) Orthorhombic symmetry; (c), (d) monoclinic symmetry; (e) triclinic symmetry. The orthogonal axes a, b and c are those used by Sanders (1926) to describe the geometry of symmetrical cylindrical folds. b coincides with the hinge line of the fold (sometimes referred to as B), c is at right-angles to the axial plane and a is at right-angles to b and c.

structure. As a rule, the bigger the structural picture that one can see, the more accurate is one's tectonic interpretation.

Several systems of co-ordinates or structural axes have been used by various workers to facilitate descriptions and comparisons of orientations of rock-structures. These have been concisely summarised by Ernst Cloos in his memoir on *Lineation* (1946, p. 5). The terminology, which is now in most general use, is that of Sander (Sander & Schmidegg, 1926) developed during his work on tectonite fabrics; it is in many respects the same as that proposed by Jannetaz (1884). Heim (1921, p. 83) used the same axial directions, but gave them different letters. Considerable confusion and misunderstanding have arisen concerning the correct use and meaning of Sander's axes of symmetry. The reason for this confusion is discussed by Siddans (1972). He points out that the essence of Sander's method for studying tectonite fabrics is the analysis of geometric relations of all measurable elements in a rock – crystallographic orientations of individual grains, grain shapes, and the arrangement of particular kinds of grains into some geometric configuration. From the geometric analysis a kinematic interpretation is made in terms of translation, rotation and strains, i.e. the 'movement picture' of Sander. Dynamic interpretation is then made in terms of body forces and surface forces. These steps are made with increasing uncertainty. In deriving the kinematic from the geometric analysis, the key factor advocated by Sander is symmetry, the concept being that the symmetry of the movement picture is reflected in the symmetry revealed by the geometric analysis. The dynamic analysis follows from assumed constitutive laws (stress–strain relationships).

Sanders used symmetry classes and **orthogonal fabric axes** a, b, c to emphasise the dominant structural features of the rocks. Thus 'spherical fabrics' have randomly oriented fabric elements; a planar fabric is the ab-plane, with any conspicuous linear element in the plane defining the b axis. Planar fabrics may be of four types: axial, orthorhombic, monoclinic or triclinic, according to the relationship of the various fabric-element sub-fabrics (see Turner & Weiss 1963, Fig. 4.10).

Sander also used orthorhombic **kinematic** axes, a, b, c which are only meaningful in simple shear (see Fig. 3.2) and **orthogonal** axes a, b, c to describe fold geometry which are only meaningful for cylindrical folds (Fig. 3.1a). The presentation of three sets of orthogonal axes, identically labelled yet quite independent in their meaning provided a situation fraught with confusion and with considerable scope for abuse. Kinematic axes were equated with finite strain ellipsoid axes, previously denoted by a, b, c (Heim 1921), a totally invalid practice in most geological deformation (see Ramsay 1967, Fig. 6.45); and ill-conceived references

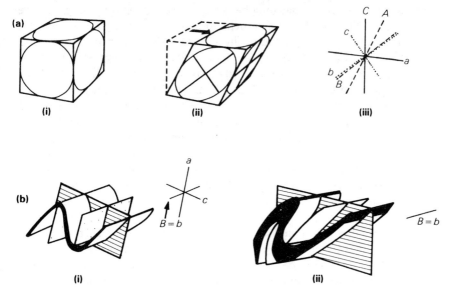

Figure 3.2 (a) (i), (ii) Simple shear deformation. (a) (iii) shows the kinematic axes, a, b, c, and the axes of the finite strain ellipsoid A, B, C. Axes, a, b and c are mutually perpendicular as are axes A, B and C. B and b are parallel. (b) Cylindrical folds referred to orthogonal axes a, b, c. The b axis is parallel to the fold axis B. The a–c plane is the symmetry plane normal to the fold axis. The a- and c-axes are only uniquely defined when the axial surface of the fold is planar ((a) & (b) (i) after Siddans 1972, Fig. 22, (b) (ii) after Whitten 1966, Fig. 8b).

to 'movement', 'flow' and 'tectonic transport' based on fabric or fold axes, and thought to define the kinematic axes in situations where they are quite indefinable, proliferated in the literature.

It is apparent from the above discussion that the steps leading from a geometric analysis to a kinematic interpretation and then to a dynamic interpretation should be taken with great care. It is generally incorrect to assume that Sander's fabric (i.e. geometric) axes a, b, c coincide with the maximum, intermediate and minimum extension directions of the strain ellipsoid and that these coincide in orientation with the minimum, intermediate and maximum compression direction of the stress ellipsoid.

However, the geometry of certain structures does seem to indicate certain features about their kinematic development and the stress fields associated with them. For example, as a general rule, the presence of well-defined monoclinic structures, or of recumbent orthorhombic structures, indicates the action of a shearing couple on the rocks in question.

The principle that a monoclinic structure indicates a movement of monoclinic symmetry has long been tacitly recognised by geologists. The very term *déjeté* used by the French, and applied to an asymmetrical fold, suggests movement or bending in a certain direction; and Thurmann (1853, quoted by de Margerie & Heim 1888, p. 110) suggested the term *régard d'un pli* for the direction towards which the steeper flank faced. As pointed out by Fallot (1957) this term has the same significance as the German word *Vergenz*, and both imply a sense of direction. Sir Edward Bailey (1935) defined the direction of movement, as understood in tectonics, as 'the direction of relative movement of the more superficial layers of the Earth's crust', and he explains his definition by referring to the direction of horizontal travel of the crest of an overturned anticline, relative to its root. This movement will normally be at right angles to the axis or direction of the hinge of the fold. But one minor fold axis, and one *régard* do not define a regional tectonic direction any more than one swallow makes a summer. It is only by noting *all* the structures in the field and understanding their significance that a true interpretation of the tectonic pattern and history can be obtained, and by far the most important of these minor structures are the axes of the major and minor folds of the region, that is to say, the *b* or *B* structural axes.

The term **axis of a fold** has been defined differently by different authorities, and many have considered that it is synonymous with the crest or trough of a fold. Billings (1942), Moret (1947) and Nevin (1949) defined it as the line of intersection of the stratification with the axial plane. Haug (1927, I, p. 194), Bonte (1953) and Stockwell (1950) considered the axis as the hinge-line of a fold, or the line along a particular bed where the curvature is greatest. Clarke and McIntyre (1951), discussing these definitions, pointed out that many complex folds maintain remarkably constant profiles for considerable distances along their lengths. They agreed with Sander (1948, p. 59) that the shapes of such folds are open cylindrical structures which could be formed by a generatrix moving parallel to itself and the axis of the cylinder. Hence, following Wegmann (1929), they stated that the axis of a fold can be defined as the nearest approximation to the line, which, moved parallel to itself in space, generates the fold. This definition was based 'on the usage of the Alpine structural geologists (especially Lugeon and Argand) since the 1890s'. A complex fold system may thus be pictured as a surface (or bed) partially wrapped around a bundle of parallel cylinders. The direction and inclination of the structural axis of the whole system is not necessarily confined to any particular horizon, but is declared by any straight line drawn on any one of the cylindrical surfaces (Fig. 3.3). The crests, troughs and hinges are, therefore, but special positions of the generatrix, which itself defines the fold-plunge.

Structural symmetry

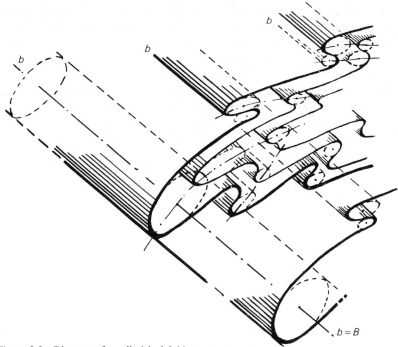

Figure 3.3 Diagram of a cylindrical fold system.

The normal cylindrical fold has but one plane of symmetry, at right angles to its axis; and its axial direction, $b = B$, will be common to both large- and small-scale structures. Hence by noting the b-axes of the small-scale structures we should be able to define the orientation of the main structure.

It has, however, been pointed out by Goguel (1952, p. 126) that '*il arrive souvent que la surface du banc-repère n'affecte pas de tout une forme cylindrique*',* and, as shown by Stockwell (1950), an anticline, which along its length dies out and changes into a syncline, may approximate to a conical structure. Such non-cylindrical folds will declare themselves by the non-parallelism of the b-axes of the minor folds which are dependent on them, provided enough observations are made.

As pointed out also by Ramsay (1960), many shear folds, though at first sight monoclinic, are in reality triclinic. He observed that the symmetry axes of shear folds and the symmetry axes of associated mineral fabrics are often not coaxial.

* . . . 'one commonly finds that the surface of a marker bed is not at all cylindrical'.

Admittedly many large-scale structures are not truly cylindrical; but among them there are many which are not far off being so. Also many structures are approximately cylindrical over short distances along their lengths. Hence, if one divides a region into subsidiary areas, and considers both the axes and the symmetry of the structures in those areas, at first individually, and then in conjunction or comparison with those in adjacent areas, one can often arrive at a composite picture of the tectonics of the region as a whole. The method was outlined by Wegmann (1929), and has been successfully used in the elucidation of superposed structures in the Highlands of Scotland by Johnson (1957); Ramsay (1958a); Clifford *et al.* (1957); Clifford (1960) and Treagus (1974).

4 Structure and stratigraphical succession

One of the most valuable uses of small-scale structures to the 'field geologist', is the manner in which he can employ them to determine the local stratigraphical succession. This, however, is only possible within certain limits; and determinations of stratigraphical order, made on the evidence of secondary structures, must always be considered as *first approximations* which may be liable to modification as the work progresses.

The small-scale structures which can be used in determining the stratigraphical succession can be subdivided into two groups: (a) original sedimentary structures, and (b) secondary structures of tectonic origin; both are discussed by Shrock (1948).

The original sedimentary structures are those which were formed as direct results of sedimentation processes, and, if clearly preserved, they yield definite evidence of the top and bottom of individual beds. The structures include such phenomena as graded bedding, current bedding,* ripple marks, and sole markings, etc.

It is the secondary structures of a tectonic origin that can be used to determine stratigraphical successions which are discussed in this chapter; they include tension gashes, cleavage and schistosity, and drag-folding,

---

* Current bedding was used to determine the stratigraphical order of beds in Ireland before 1864 (Lamont 1940). Its use was known to some geologists in the United States early in this century, and A. C. Lawson used it in the 'Rainy Lake' area of Canada in 1913. He acknowledged that W. O. Hotchkiss had previously demonstrated its significance. In Great Britain the importance of the structure was forgotten; and, despite the fact that it was used by J. F. N. Green (1924) and J. J. Hartley (1925), its significance was not appreciated until 1930, when Sir E. B. Bailey's attention was drawn to it in the Ballachulish area of Scotland (Vogt 1930, Tanton 1930, Bailey 1930, 1934). Since then it has been used extensively in elucidating the structure of the metamorphosed rocks of the Scottish Highlands and elsewhere, (Richey & Kennedy 1939, Sutton & Watson 1955, Read 1958). To refer to the stratigraphical significance of current bedding as *Le Loi de Bailey* (Bonte 1953) seems to me to ignore the normal rules of precedence.

etc., and the evidence they present will be discussed under each appropriate section. Unfortunately many of these small-scale tectonic structures may be produced in a variety of ways. In consequence their interpretation must always be considered tentative until the weight of evidence confirms (or definitely disproves) the original explanation. If the structures observed were formed during more than one tectonic phase, then the evidence they present may be misleading; unless one can distinguish between those small-scale structures which are related to different tectonic phases or episodes.

It is only in regions where a single phase of folding has been the dominant tectonic movement that small-scale structures can be safely used to determine the correct local stratigraphical succession. If the major folding and the minor structures associated with it, have all been produced by one and the same general tectonic movement, then the minor structures will largely result from the relative slip between the folded strata, developed while the major structures were growing.

When beds are folded by flexure, there is always a tendency for the *upper strata to slip upwards over the lower strata towards the hinges of anticlines, and away from the troughs of synclines* (Leith 1923, p. 176) (Fig. 4.1). This slip may occur between individual beds, or between groups of beds; and, if the strata at the beginning of the folding episode were in their correct stratigraphical order, then even if the fold is overturned, it is only necessary to observe which beds *have moved upwards* relative to their neighbours, to decide which are the younger and which the older.

This slip between the beds can easily be demonstrated by folding a thick book, or sheets of cardboard or felt, and observing the relative movement between adjacent sheets. Kuenen and de Sitter (1938) also found experimentally that, when a homogeneous slab of plastic clay was folded, shearing planes which were concentric with the top and bottom of the slab made their appearance. The movement on these planes was upwards towards the anticlinal crests. Evidence that such slip does occur between beds which have been folded was recognised by H. Cloos and Martin (1932) in the field by the displacement of pre-tectonic quartz veinlets. Individual veinlets were displaced up the dip by movement on the planes of stratification where they passed from one bed to the next. Similar displacements of gold-bearing quartz veins, which were offset up to 60 cm between individual beds, have been described by Kenny (1936) and Hills (1945) (Fig. 4.2a).

This slip may result in the polishing or grooving of the bedding planes (de la Beche 1853, Lewis 1946); and Nieuwenkamp (1928) noted that the angle between the dip of the strata on the flanks of a fold and the plunge

Structure and stratigraphical succession 19

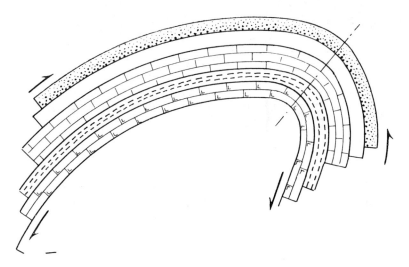

Figure 4.1 The upward slip between beds on a flexure fold.

of slickenside striations formed during the folding (α) was equivalent to the plunge of the major fold itself (ϕ) (Fig. 4.2b). Wegmann and Schaer (1957) observed in the Jura, that, in addition to slickenside striations, 'crescentic grooves', which they termed *lunules*, may also be formed by this same mechanism. In shallow depth folding, such as they have in the Jura, it was found that the movements indicated by the striations and *lunules* were not always parallel, even on adjacent planes of stratification. This means that, in this case, the slip between the beds was not always directly at right angles to the present hinge of the fold, and that the movements had been influenced by lateral adjustments during the development of the flexures.

In a detailed investigation of the minor structures associated with a fold system in the Upper Viséan limestones near Boeyberg (Plombières), de Waard (1955) noted that the striations observed on bedding planes were strongly dispersed. The orientations of the striations were, however, symmetrically scattered on either side of the line normal to the fold hinge. The effect of this interstratal slip is to generate a stress-couple which acts on the upper and lower surfaces of individual beds or groups of beds. In each fold, the torsional sense of the couple will be towards the anticlinal hinge, and so will be opposed on opposite flanks of the same fold (Figs 3.2, 4.3). Hence, although the major movement which gives rise to the folding may have unique sense, the subsidiary movements and torsional stresses within the folded beds themselves will be opposed, and will form monoclinic structures which will declare the direction of local

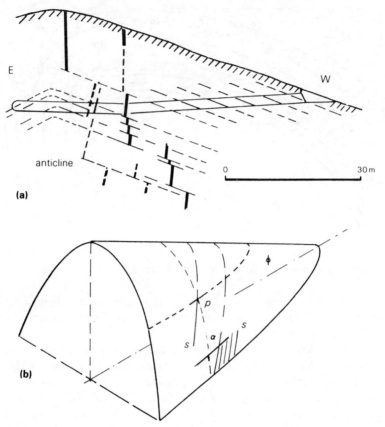

Figure 4.2 (a) The displacement of gold-bearing quartz veins in the 'Golden Stair Mine' Australia, up the planes of stratification, towards the anticlinal crest (after Kenny 1936). (b) The relationship of the orientation of slickenslide striations (s) to the fold-plunge (ϕ) (p = dip of bedding; α = angle between the striations and the direction of dip). (After Neuwenkamp 1928, Nevin 1936.)

slip. In a folded terrain, reversal in the local direction of movement, as shown by the minor structures, commonly indicates the crossing from one flank of a fold to another. The relationship between the relative sense of movement of strata and the stratigraphical succession is shown in Figure 4.3 in which the arrows indicate the directions of slip.

The recognition of the age relationship between individual strata also gives an indication of the positions of the main structural units. Since the core of an anticline is composed of older rocks than is the envelope, the sense of the slip, which assists us to determine the relative ages of the beds, will also indicate whether the anticlinal or synclinal axial plane (or core) lies to the left or the right of the exposure being considered.

It is important to note that in regions characterised by recumbent folds, or by *nappes de recouvrement*, in which the axial planes and the frontal or root hinges are horizontal, it is impossible to decide, on the evidence of small-scale tectonic structures only, whether the folds which close in one direction or in the other are anticlines or synclines. The slip between the beds may be definite, and in accord with the geometry of the major structures, but the stratigraphical interpretation would be ambiguous (Fig. 4.4a). In the case of a nappe with a plunging browfold (*tête plongeante*), or of an inverted syncline (*faux synclinal*) which closes upwards, such as that seen in the Axen Nappe in the neighbourhood of the Axenstrasse, Lac des Quartres Cantons, Switzerland, the small-scale structures formed by bedding-plane slip would again declare the *geometry* of the structure, but not the relative ages of the core and envelope rocks. It has been suggested that structures which have the geometrical forms of anticlines and synclines, but in which the relative

Figure 4.3 The stress-couples developed by the internal slip between beds when folded: *a*, simply folded beds; *b*, vertical beds; and *c*, overturned beds.

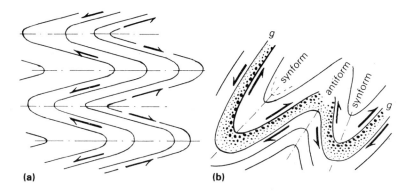

Figure 4.4 (a) Profiles of folds with horizontal planes, showing the directions of slip on the planes of stratification, towards the fold-hinges. (b) Antiforms and synforms, showing directions of slip on the planes of stratification. The fact that the fold is inverted is shown by the graded bed (*g*), and not by the minor structures.

ages of the rocks are unknown, should be referred to as **antiforms** or **synforms** until the age relationships of the cores and envelope-rocks were established (Bailey & McCallien 1937, p. 81). Where such complexities arise the problem cannot be solved by minor structures alone; stratigraphy, or sedimentary structures, which would show the direction in which the tops of the bed faced, are essential (Fig. 4.4b). This problem has been discussed by Shackleton (1958).

In a terrain of folded crystalline schists it may be difficult, if not impossible to differentiate anticlines from synclines by means of the dip of the stratification; though folding may be recognised by the repetition and local convergence or divergence of certain rock types (Fig. 4.5). If, however, the direction of plunge B can be recognised by either direct observation of the plunge of minor folds or of lineations parallel to the major fold (shown by arrows b in the figure), the zig-zags can be divided into those which converge in the direction of the plunge, and those which diverge. The former will be antiforms, the latter synforms.

If one is satisfied that large-scale inversion of the rock-sequence has not taken place, the terms anticlines and synclines can be used. The rocks in the cores of the former will then be older than those which occur on the flanks; similarly, the rocks of the cores of synclines are younger than those which surround them.

The problem which now faces us is to interpret the minor structures, which we observe in the field, into terms of rock movement or directions of forces; and to apply the results to the principles discussed above, as

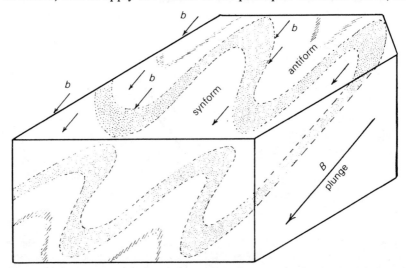

Figure 4.5 Block diagram illustrating the relationships between the horizontal plan of folds, linear structures in b, and fold plunge, B.

Structure and stratigraphical succession

far as possible. Once we have acquired an understanding of how the small-scale structures are formed, what they mean, and to what extent their significance is limited, we can apply them as aids either in the elucidation of the stratigraphy of the rocks, or in explaining the geometry, kinematics and tectonic history of the larger-scale structures. In the chapters which follow, the main types of minor structures and their significance from both a tectonic and a stratigraphic point of view are discussed. Admittedly ambiguities will arise, and lacunae in the explanations will occur; commonly we shall find that similar-looking structures may be formed by different external mechanisms. But it is hoped, that if the general forms of the more important small-scale structures are clearly recognised, and their relationships to the rock-movements is understood, the 'field geologist' will find himself armed with a useful weapon which can be wielded by stratigraphers, petrologists and tectonicians alike.

5 Structures in brittle rocks

Tension fractures and shear zones

The manner in which small-scale structures develop in the more resistant or brittle rocks, which are not appreciably deformed before rupture, is analogous to the behaviour of brittle materials — cast iron, concrete, etc. — when tested in the laboratory. The rock fails, when compressed, either by tension or by shear. Although rocks are much weaker in tension than they are in shear, we find that in the field they have failed as often by shearing as by tension. Failure by tension necessitates extension, and this may be inhibited by the containing pressure; whereas failure by shear can occur without appreciable change in volume. In many cases the two styles of rupture are found together, tensional fractures being supplementary to zones of shearing and vice versa.

Laboratory experiments have shown that a specimen subjected to a single compressional force will rupture either by tension fractures parallel to the direction of application of the force, or on two inclined planes of shear which will develop at angles of *less than 45°* to that direction (Fig. 5.1a and b). If, however, the specimen is submitted to a triaxial compression test, as would be the case of a rock buried at some depth, it is found that the orientations of the planes of fracture are controlled by the directions and relative magnitudes of the three principal stresses. 'Mathematical analysis supplemented by experimental evidence indicates that the factor which decided whether fracture would take place or not is the *difference between the greatest and the least principal stress*', $(P_{max} - P_{min})$. 'This difference was named the stress-difference by G. H. Darwin' (Briggs 1927, pp. 2–3). This means that the intermediate stress, P_{int}, can within reason be considered neutral (Fig. 5.1a). Its direction will accord with the line of intersection of the two shear planes, and with their intersection with the tension fracture surfaces. The direction of maximum compressive stress bisects the acute angle between the two shear planes, and is at right angles to this line (Anderson 1951, pp. 9–11, Hubbert 1951, E. Cloos 1955, de Sitter 1956, p. 27).

Tension fractures which result from simple compression may appear in

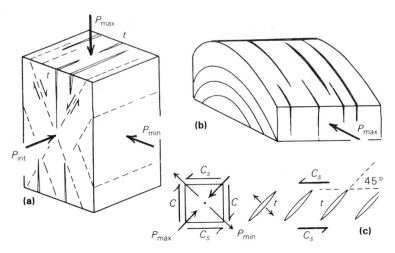

Figure 5.1 (a) The orientations of tension fractures (t) and shear fractures in relation to the three principal stresses. (b) Tension fractures parallel to the principal compressive stress and at right angles to the axis of folding. Fractures and joints in ac. (c) The resolution of stress-couple (Cs, Cs) into principal stresses, and the orientation of tension fracture (t) parallel to P_{max}.

the field as cross-joints at right angles to the axes of folds, or they may form lenticular veins filled with quartz, calcite or ore minerals. Dykes of igneous rocks may also lie along tension fractures. The veins tend to be parallel and to occur in irregular groups (Fig. 5.1b) or in zones, the trends of which are normal to the lengths of the veins, and are commonly confined to a particular structure or type of rock (Fig. 5.2).

Probably the best large-scale geological examples of the development of shear planes by compression are tear- or wrench-faults which cut obliquely across folded terrains. Those which occur in Pembrokeshire, South Wales, have been analysed by Anderson (1951, pp. 60–64), who demonstrated that, almost throughout the area, the rocks forming the acute-angled wedges between the faults have moved inwards, in a direction corresponding to that indicated in Figure 5.1a. The faults lie in two directions and are symmetrically oriented in respect to the direction of the local folds and thrusts; they closely correspond to the theoretical directions of shear. The direction of compression as indicated by the trends of the folding bisects the acute angle between the faults. The relationship between principal stresses, directions of shear and of faulting has also been discussed theoretically and in experiment by Hubbert (1951) and E. Cloos (1955) whose conclusions can be applied to both large- and small-scale structures. Small-scale shear planes may show in exposure as small faults or as thin intersecting veins; but commonly an

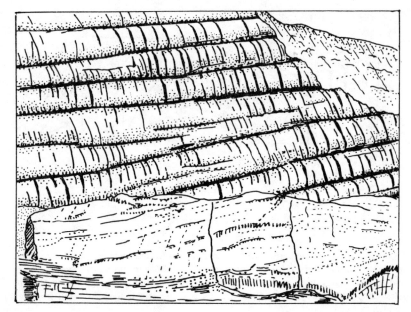

Figure 5.2 Tension fractures cutting the axes of minor folds, at Rillage Point, North Devon (after A. J. Bull 1922).

incipient fault plane is replaced by a zone of tension gashes *en échelon*. The formation of such tension gashes results from the yielding of the rock within a narrow zone in the immediate vicinity of the incipient plane of shearing (Fig. 5.1c). The shear stresses (Cs) which, to prevent internal rotation must be balanced by an induced equal and opposite shearing couple (c), can be resolved into a compression (P_{max}) in one direction and a relative tension (P_{min}) at right angles to it. If the stress-difference is great enough, tension gashes (t) (oriented at 45° to the plane of shearing) are formed along the zone of shearing (Fig. 5.1c)*. Goguel (1952, p. 50, 1953) considered that, if the rock contains a pore-fluid under high pressure, the hydrostatic pressure of the fluid would assist in forcing the walls of the fractures apart, thus reducing the stress-difference that might be necessary to cause the walls of the fracture to gape.

Fractures showing the same *en échelon* pattern as those discussed

---

* This mechanism, on a large scale, has been invoked to explain the long zones of faults *en échelon* which occur in Oklahoma, Montana and Wyoming and also on the Mexia Fault in Texas. It is considered that horizontal movements on faults in the basement have formed zones of normal or tension faults tens of kilometres long in the cover rocks above, Fath (1920); for other views see Sherrill (1929), Melton (1929) and discussion in de Sitter (1956, p. 173).

Structures in brittle rocks 27

above, were noted by Gilbert (1928) in unconsolidated sediments along the line of the San Andreas Fault. Similar fractures, sometimes referred to as 'feather joints', also occur in the walls of faults, and on the margins of igneous intrusions (E. Cloos 1932, Balk 1937, p. 102, Fig. 35, Wilson 1951, Fig. 13a, p. 424 & 1960).

Single zones of tension gashes *en échelon* are good examples of a monoclinic structure but a conjugate pair of such zones as shown in Figure 3.1b, forms a structure showing typical orthorhombic symmetry.

The three symmetry axes of the conjugate pair of shear zones are parallel to the three principal stresses responsible for the formation of the shear zones. The *maximum* compression P_{max} bisects the acute angle between the two shear zones (Fig. 3.1b), the *intermediate* principal stress is parallel to the intersection of the two zones and the *least* principal stress bisects the obtuse angle between the two zones.

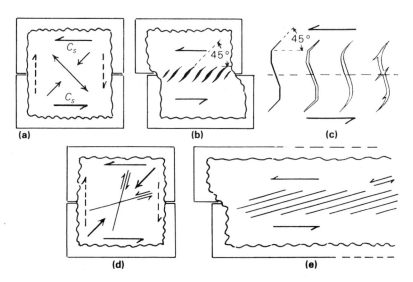

Figure 5.3 Riedel's experiment. (a) The stress distribution; (b), (c) the development and curvature of tension fractures; (d), (e) the formation of shear fractures and their orientation with respect to the principal line of fracture and movement.

Tension fractures *en échelon* have been reproduced experimentally by Riedel (1929) in a slab of wet clay resting on two boards (Fig. 5.3). The whole suggests a formation of sediments overlying a rigid basement cut by a single fault here represented by the junction between the two boards. A slight horizontal movement of one board relative to the other causes fractures to appear in the clay at 45° to the line of slip, as shown in Figure 5.3a and b. With a little more movement, the acute ends of the fractures, where the stresses are concentrated, are extended further in the 45° line; but their central portions and the clay between them tend to rotate. The fractures thus assume a sigmoidal form (Fig. 5.3), as was illustrated by Shainin (1950). This shape may become even further developed, and secondary or subsidiary fractures may appear. In some cases the ends of the main fractures may merge into planes of shear which are nearly parallel to the direction of shearing (Wilson 1960).

The development of naturally occurring fractures *en échelon* has been likened by many workers to those produced experimentally by Riedel. Such structures are commonly associated with tectonic shear zones such as those observed and recorded by Skempton (1966); others on a larger scale have been illustrated by Tchalenko and Ambrasis (1970) from the Dasht-e Bayez fault zone in Iran, which was responsible for the 31 August 1961 earthquake in Iran. The movement on the main zone of displacement was dominantly left-handed and horizontal. A diagram showing a simplified picture of the pattern of the fractures is shown in Figure 5.4 based on Skempton (1966) and Wilson (1970). The lettering on the different fracture planes observed corresponds to that proposed by Skempton, R and R' are Riedel shears, that is, those obtained by Riedel in his experiments. D are shear fractures (Skempton's displacement shears) parallel to the principal shear plane or zone of displacement; T is the direction in which tension gashes may form. These may curve and continue as shear fractures, similarly shear fractures may curve and continue as tension fractures.

Innumerable examples of tension gashes *en échelon* may be seen in the *calcaire bleu* of Belgium. Shainin (1950) has also pointed out that if the

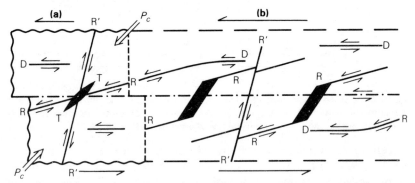

Figure 5.4 (a) The relationships of the fracture patterns to the orientations of the principal stresses in a wrench-zone developed as the result of left-handed simple shear. (b) Generalised plan-view of the fracture patterns that can develop above, within, or in the vicinity of a zone of left-handed wrench movement. P_c, principal compressive stress; R & R'; Riedel shears; D, displacement shears; T, tension fractures (after Wilson 1970).

Structures in brittle rocks

shearing movement continued as a plastic deformation of the rock mass as a whole, an ambiguous structure may develop. In this case it is possible that the ends of the fractures may have been dragged around so that they 'candle-flame' or curve in the direction of the main movement instead of in the opposite sense.

Intersecting zones of tension fractures *en échelon* mark the incipient development of the two directions of shearing, and were termed **conjugate sets or zones** (Fig. 3.1b) by Shainin (1950). The structures which this author described and figured occur in argillaceous limestone, and the veinlets are filled with calcite. He noted that the zones were nearly vertical, and that the acute bisectrix of the angle between pairs of zones was normal to the trends of the local folds, that is, parallel to the local maximum compressive stress. The orientations of these structures,

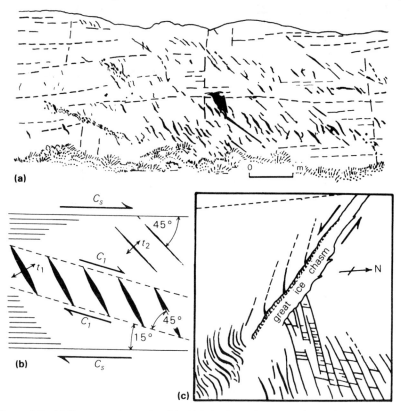

Figure 5.5 (a) Quartz tension gashes (black) in horizontal quartz–mica schist; (b) analysis of the fracture patterns shown in (a), and of the probable stress distribution responsible (after Wilson 1953); (c) a tracing of an oblique aerial view of the 'Great Ice Chasm' Filchner Ice Shelf, Antarctica, and its crevasses (after Wilson 1960).

and the stress-distribution responsible for them, are thus comparable with those found in tear-faults.

Similar structures have been observed in the Torridonian Sandstones near Kyle of Lochalsh, in NW Scotland. The Sandstones now form a big recumbent syncline which was overturned by the Moine Thrust movements; its axis trends NNE. The tension gashes vary from thin films, to veins 10 cm thick, and from 3 cm to over 1 m broad. They are mostly filled with quartz, but some contain pegmatite. Kanungo (1956) plotted the intersections of the various zones, and determined accurately the lines along which they cut each other, these closely accorded with the observed direction of the B axis (hinge) of the major fold. He also determined the local orientations of the maximum and minimum stresses. The manner in which these latter orientations changed as the fold was traversed normal to its axis, suggested to him that the conjugate zones had developed during the period when the recumbent syncline was actually closing.

Riedel (1929) also found in his experiments that in addition to tension fractures he could produce fractures formed by shear. These were oriented at about 15–20° to the surface of slip between the two boards (Fig. 5.3d and e). Ernst Cloos (1955), using a slightly different technique, obtained both sets of shear fractures: one was nearly parallel to the direction of slip and the other nearly at right angles to it.

A large-scale example of the development of fractures formed by shear and by tension has been observed on the Filchner Ice Shelf, Antarctica (Fig. 5.5c) (Fuchs & Hillary 1958, Wilson 1960). Here in the vicinity of a big zone of horizontal movement – the 'Great Ice Chasm' – the ice is cut by crevasses, the pattern and forms of which closely accord with the results obtained by Riedel and E. Cloos. To the south of the east end of the Chasm are sigmoidal tension gashes, to the north are crevasses which correspond to the two directions of maximum shear.

Closely-spaced planes of shear, which corresponded to those obtained experimentally by Riedel, have been observed by Blyth (1950) in some porphyrite dykes which had suffered shearing parallel to their margins. Wilson (1952) described an exposure of the Moine Series in Scotland, in which two groups of tensional veins had been formed as the result of a nearly horizontal shearing movement along the local plane of schistosity (Fig. 5.5a and b). One group of veins (t_1) lay in zones which make an angle of about 15° to the schistosity, and these marked zones of incipient shearing (C_1). The second group (t_2) comprised veins which make an angle of 45° to the schistosity. These latter had been formed by a tensional stress which had acted throughout the mass of the rock as a whole. The generating force responsible for the structure, from evidence found elsewhere in the area, was a slip, C_sC_s, parallel to the foliation, and at right-angles to the elongations of the quartz veins, this also agreed with the stress distribution that can be deduced from the structure itself.

Slip along bedding planes during folding may also produce structures similar to those described above, and these can, in some circumstances, be used to determine whether the strata are in their correct order or are inverted. The frictional drag which results from the upward slip of the upper beds over the lower, towards anticlinal hinges, imposes a shearing couple which acts along the top and bottom of each individual bed. If conditions are favourable, tension gashes at 45° to the stratification will develop (Fig. 5.6). The orientations of the fractures in relation to the stratification will declare the direction of slip.

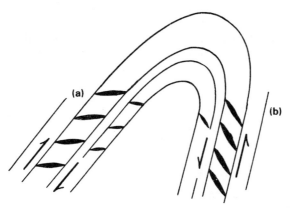

Figure 5.6 The development of tension gashes in fold beds: (a) beds right way up; (b) beds inverted.

In all the examples of the fracturing of brittle rocks discussed so far in this section, it will be noticed that both tension fractures *en échelon* and shear fractures face, as one might say, against the direction of shear movement. With a prolonged movement therefore, the thin strips of rock between the planes of fracture may not be able to withstand the torsional stresses, and they will break. Commonly the central portion of a zone of tension gashes *en échelon* has been rotated and shattered, with the formation of a breccia zone, from the sides of which the ends of tapering gashes project. Examples can be seen in the polished marble slabs used in shop-fronts in many cities.

The development of zones of closely spaced parallel fractures, which may even grade into a variety of fracture cleavage, can also be recognised as being related to the directions of maximum shearing stress in the rocks containing them. In the early stages of the application of tectonic forces, the strata yield by folding; and as the movement continues they may even be rendered schistose. As the limit of relief by folding is approached, the rock-mass as a whole becomes more and more rigid, and if recrystallisation occurs this rigidity may be still further increased. The folded strata will now behave, not as a stratified pile of beds, but as a homogeneous block, which, if the compression is sufficiently great, will yield by fracture parallel to the local directions of maximum shearing stress. Such fractures, formed by the same stress-system that was responsible for the folding, tend to be symmetrically oriented with respect to the fold axial planes and to the schistosity, and these two are normally parallel. The resulting structures and their mode of formation were described and discussed by Muff (Maufe) (in Peach, Kynaston & Muff 1909). Similar structures have been recognised by Mead (1940) who agreed with Muff's interpretation of these fractures and suggested that fracture systems formed in this way should be termed **shear cleavage**. Muff, in a simple diagram, illustrated the relationship between the principal compressive stress, the folding, the axial-plane cleavage or

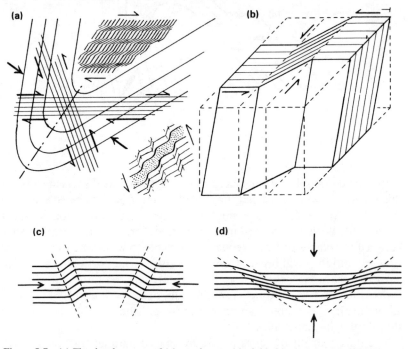

Figure 5.7 (a) The development of 'shear cleavage' in folded beds (after Muff 1909), and showing the manner in which stratification and schistosity may be affected by it. (b) Joint-drags (based on Flinn 1952, Fig. 2, p. 266). (c) Conjugate 'reverse' kink-bands. (d) Conjugate 'normal' kink-bands (full arrows represent the principal compression direction).

schistosity, and the orientation of the later cleavage planes (Fig. 5.7a). The sense of slip on the later fractures is indicated in the diagram, and can be seen in the rocks themselves. It is not common for both sets of fractures to develop together; more often one set will appear in one limb of the fold, and the second set in the other; or as Mead noted, only one set of fractures may appear, with the exclusion of the other.

In the field the structure may be seen as zones of mechanical cleavage in which the gliding planes are spaced from 0.25 to 0.5 mm (Muff 1909). The result is a rucking of the earlier schistosity and of thin strata, and the slicing of the more competent beds by a fracture cleavage. Otherwise good roofing slates are often rendered completely useless in an area where they are traversed by these zones of secondary cleavage. The original schistosity (slaty cleavage) is bent or dragged into the planes of slip, along which secondary mica or chlorite commonly crystallizes.

Phyllites and crystalline schists as well as slates may be traversed by these zones which may or may not be fully developed. Such incipient fractures are represented by monoclinal wrinkles on the schistosity surfaces. These wrinkles form a marked linear structure and may be closely spaced, or they may be several centimetres or even a metre or more apart. In profile, they can be seen to traverse the schistosity in subparallel zones which gradually disappear in one place and then start again elsewhere *en échelon*.

The relationship of the conjugate cleavage planes to folds described by Muff is summarised in Figure 5.7 which shows the compression responsible for folding bisecting the acute angle between the cleavage planes. Detailed measurements by Knill show that this angle is generally well in excess of 90°. An alternative mechanism for the formation of the conjugate cleavage has been proposed by Cosgrove (1976) who suggests that they are conjugate normal kink bands probably caused by the continued action of buckling stress on the well developed axial-plane cleavage.

Kink bands are monoclinal flexures that develop in materials with a good mechanical anisotropy. They have been classified by Dewey (1965) into normal kink bands and reverse kink bands. The formation of these structures has been discussed by Cobbold *et al.* (1971) who show that reverse kink bands form when the principal compression is parallel or at a low angle to the layering or fabric in a rock (Fig. 5.7c) and normal kink bands form when the principal compression is normal or at a high angle to the layering or fabric (Fig. 5.7d). Kink bands may develop as conjugate sets and where they do so, the principal compression always bisects the *obtuse* angle between them (Fig. 5.7c and d).

Examples of kink bands have been illustrated by Dale (1899) from the 'slate belt of Eastern New York and Western Vermont'. He termed them simply 'zones of shearing', but the local quarrymen called them 'hogbacks'. Read (1934) recorded similar structures in the Muness Phyllites of Unst in the Shetland Islands. These have been re-examined by Flinn (1952) who discussed their formation, and suggested that they should be referred to as joint-drags (Fig. 5.7b). Flinn observed that each individual unit had a width of about 2.5 cm; elsewhere I have seen them cutting thin bedded siliceous and semipelitic metasediments of the Moine Series, and having widths of 15–20 cm.

Other examples have been illustrated and discussed by Hoeppener (1956). Reverse kink bands can be reproduced in a test-piece of hard wood if it is compressed to the point of failure in a direction parallel to that of the grain and in well cleaved rocks if compressed parallel to the cleavage (see Patterson & Weiss 1968).

The orientation of kink bands may be related to larger fold structures, if, for example, they are caused by the same stress field. However, it is quite possible that these structures, when observed in the field, may have been the result of some later tectonic movement which was entirely unrelated to the earlier folding episode. In these cases, though the structures appear as extra complications, they may be very helpful as guides indicating the sequence of movements and the later positions of the principal stresses which were responsible for them.

6 Rock cleavage and schistosity

Generalities

Rock cleavage, and here I include schistosity, as a structure by which rocks are fractured or cleaved into thin slices distinct from stratification, appears to have been first recognised by geologists about 1820. It was observed by MacCullach (1819) in the rocks of the Western Isles of Scotland and by Otley (1820, 1823) in the English Lake District. Adam Sedgwick (1835) must be credited with the first truly scientific study of the phenomenon; but more thorough investigations of cleavage and schistosity, and theories of their origin were only seriously started about the middle of the last century. Prominent among the early workers were Beete Jukes (1842), Sharpe (1847, 1849), Darwin (1846), Sorby (1853, 1855, 1856), Laugel (1855), Haughton (1856), Phillips (1856) and Tyndall (1856). The early investigations into the problem were surveyed by Phillips in 1856, and later, when more material was available, by Harker in 1885. E. Cloos (1946, p. 58) considered that: 'This paper (by Harker) is fundamental . . .' The annotated bibliography in the memoir on *Lineation* by Cloos (1946) contains many valuable references to, and brief summaries of the early papers on rock cleavage. More recently, a most useful pictorial 'Atlas' illustrating a wide variety of examples of rock cleavage, collected by Baily, Borrodaile and Powell (1977) has been published by the University of Tasmania.

The term **cleavage** in English has a very general application. Though primarily defined as a characteristic property of minerals, it is now used, not only in connection with fracture cleavage, but also for the easy direction of splitting in roofing-slates, where in French one would refer to '*Schistosité*'.

Schistosity in English is mainly confined to the prominent plane structure of the crystalline schists and, at times, of phyllites. **Fissility** and **Fissile** are more general terms, which denote quite simply that the rock can be easily split into thin laminae. **Foliation**, on the other hand, has a definite significance in Great Britain, which has unfortunately been lost in America. It refers to a banded or ribbon structure in metamorphic rocks, in which the laminae are discrete. Harker (1932, p. 203) following

Darwin's description of the term (Darwin 1846) defined foliation as '... a more or less pronounced aggregation of particular constituent minerals of a metamorphic rock into lenticles or streaks or inconsistent bands, often very rich in some one mineral and contrasting with contiguous lenticles or streaks rich in other minerals'. Brammall used the term 'streaky bacon' structure to describe a foliated rock. Hence, according to Darwin and Harker, a foliated rock is not necessarily schistose; nor is a schistose rock necessarily foliated, but it can become so, either as a result of further deformation, or as it grades towards a gneiss, by mineral segregation or by *lit-par-lit* injection. Schistosity and foliation can commonly be distinguished in the field; and it is, in my opinion, a great pity if the two terms, both clearly defined, become used as synonyms, to provide, as Fallot would have said: another example of '... négligences dans l'emploie des mots scientifiques'.*

Varieties of cleavage have been described by several workers most of whom followed Leith (1905). More recent classifications include Leith (1923); B. and R. Willis (1923); Fourmarier (1949a, I, p. 654); Knill (1960); and others, who subdivided the cleavage types into three principal categories:

(a) **Original cleavage or parting:** stratification, fissility, etc. lamination, etc. non-tectonic.
(b) **Fracture cleavage:** false cleavage, strain-slip cleavage, etc.
(c) **Flow cleavage:** slaty cleavage, schistosity, foliation in the American sense.

Mead's 'shear cleavage' (1940) has already been considered (p. 32): it commonly appears in the field as a fracture cleavage or strain-slip cleavage.

Axial-plane cleavage is a term of non-genetic significance which can be applied to a cleavage (or schistosity) which lies parallel, or sub-parallel to the regional fold axial planes. Some authors consider it synonymous with

---

* Knopf (1941) and Fairbairn (1949) defend the American usage of the term 'foliation', tracing it back to Scrope (1823). I have not been able to see this early reference, but Scrope certainly used the term in a very broad sense. For instance, on p. 233 of his *Considerations on volcanoes* (1825), he referred to an 'increase in mica (in a granite) creating a foliate structure by the platlets being placed in a parallel position, and then the rock passes into gneiss, or foliated granite ...' On the page following he included gneiss, mica-schist and clay-slate as 'foliated rocks'. It is because of this general use of the term that Knopf and Ingerson (1938, p. 7) stated 'Foliation is used ... to denote a parallel arrangement of minerals. A foliate rock will split more or less readily into thin slabs parallel to the foliation planes. Where the rock splits into thin layers, it is called a schist'. Ragan (1967) defends this American usage in his discussion of planar- and layered-structures in glacial ice.

schistosity or flow cleavage; but fracture cleavage, especially in the vicinity of fold hinges, may also lie parallel to the fold axial planes. The term should be confined to non-genetic descriptions of the geometrical relationship of the cleavage or schistosity to the major structure.

Original cleavage, parting or fissility

This includes stratification and lamination in sediments, flow structures in igneous rocks and certain gneisses and does not greatly concern us in this study. Many rocks, such as micaceous sandstones, part or cleave into thin slabs parallel to their planes of stratification and so form flags or flagstones, in which the parting is mainly controlled by the micaceous laminae. Many clays when compacted under a load of superincumbent strata have much of their included water squeezed out, and their thicknesses decreased as the weight or thickness of superincumbent load increases. This leads to the formation of a **shale** in which is developed a horizontal fissility, or direction of easy splitting. The clay minerals are re-orientated so that they lie flat, and delicate fossils are compressed, normal to the vertical pressure and approximately parallel to the bedding.

If at a later date, such rocks become metamorphosed without tectonic re-orientation of the early layer-lattice minerals, a true mimetic bedding-plane schistosity will develop, in which the original clay minerals may have acted as nucleii or seed minerals for further growth (Knopf & Ingerson 1938, p. 39). Similarly, original variations in the compositions of the beds may, on recrystallisation, lead to the production of gneissose rocks exhibiting foliation.

Fracture cleavage

Fracture cleavage is a mechanically formed structure which is not necessarily accompanied by recrystallisation or metamorphism of the rock. It can be defined as a simple cutting of the rock into discrete, more or less thin slices, formed by a series of small parallel fractures. The cleavage planes are separated by thin sheets of rock called **microlithons** which are not cleaved. Consequently the cleavage is termed a non-pervasive cleavage. The structure referred to here as fracture cleavage has also been known by numerous other names: false cleavage (Harker 1885, Dale 1899); close joints cleavage (Sorby 1857); strain-slip cleavage (Bonney 1886); *Ausweichungsclivage* (Heim 1878, ii, p. 53); shear fracture

(Fairbairn 1949); slip cleavage (White 1949); and crenulation cleavage (Knill 1960). A summary of the early terminology is given in de Margerie and Heim (1888, p. 120); and a useful *Lexique des principeaux termes touchant a la schistosité* is contained in Baer (1956). It is evident that not all these terms refer to the same type of structure, and it has been suggested that a distinction between some of them could be made. Fracture cleavage is a general term suggesting to the reader a series of clean-cut surfaces of fracture and should be retained for a simple mechanical cleavage in otherwise homogeneous, unfoliated rocks. The rock in the thin slices (microlithons) lying between the closely spaced surfaces of cleavage, or incipient fracture, is unchanged from its pre-cleavage state, and the fracture planes themselves are independent of any parallel arrangement of the minerals which form the main rock

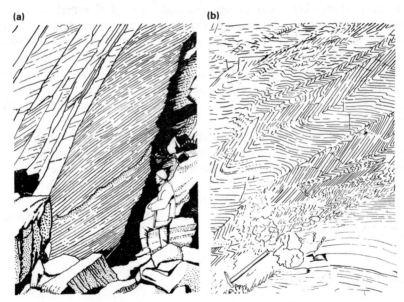

Figure 6.1 (a) Fracture cleavage in the inverted limb of a fold, South Stack Series, Anglesey, North Wales, from a photograph. (b) Coarse strain-slip cleavage cutting an earlier axial-plane schistosity, Culm Series, Boscastle, Cornwall, from a photograph (after Wilson 1951).

mass (Figs. 6.1a, 6.2). Strain-slip cleavage, false cleavage, slip cleavage, crenulation cleavage and *ausweichungsclivage* are terms which are more applicable to the closely spaced parallel zones of discontinuous fractures accompanied by small-scale plications which traverse rocks that are thinly bedded or are already foliated or schistose (Figs 6.1b, 12.3c and d). Bonney's term 'strain-slip cleavage' (1886) has been defined by

Rock cleavage and schistosity

Figure 6.2 Fracture cleavage grading to flow cleavage in greywacke and slate on the inverted limb of a fold, South Stack Series, Anglesey, North Wales.

Holmes as: 'A variety of cleavage . . . due to differential movement or "slip" along each of a nearly parallel series of closely-packed shear-planes. Between each pair of shear-planes the rocks are puckered into sigmoidal folds, the outer limbs of which merge tangentially into the shear-planes' (Holmes 1928, p. 217). It corresponds to Born's *Runzelclivage* (Born 1929); and Knill (1960) has suggested the term **crenulation cleavage** for this structure. Haug (1927, p. 231) used *pseudoclivage* for fracture cleavage and, in a drawing taken from Heim, annotated the strain-slip cleavage as *plis-failles microscopiques*. De Margerie and Heim (1888, p. 94) also mention *plis-failles* in this connection. It seems to me that a term in French for this structure would be useful, and I suggest *clivage (ou fracture) aux plis-failles*, which is descriptive and closely conforms to Haug's phrase. The adjectives 'microscopic' and 'megascopic' could then be used in addition to qualify the appearance of the structure in the field.

If metamorphism should accompany or follow the formation of a fracture cleavage, the fluids within the rock will tend to be concentrated along the fracture planes. This leads to the preferential development of minerals, chlorites and mica in particular, along and in the immediate vicinity of the cleavage surfaces which would act as 'privileged paths', as described by Read, who cites numerous examples (Balk & Barth 1936, Read 1948, p. 185, Oulianoff 1958). The influence of such planes or zones on migmatisation has also been recognised. Bosworth (1910) noted that the granite veinlets which impregnated the pelitic schists on the margin of the Ross of Mull Granite in Scotland followed both the stratification and the planes of cleavage. Larger-scale examples of very similar phenomena have been described by Wegmann and Kranck (1931) and Wegmann (1938). Metasomatism, concentrated along these privileged paths, will affect the thin slices of rock between the planes, and if sufficiently intense or accompanied by further movement, may completely obscure the mechanical origin of the structure which may become indistinguishable from flow cleavage.

Flow cleavage

Flow cleavage, slaty cleavage, true cleavage, schistosity, *Transversalschieferung* (Heim 1878) − the terms are synonymous − is a structure in which new minerals grow in the rock by recrystallisation under a tectonic control. In consequence, these minerals are flattened and elongated, so that their atomic structure and planes of cleavage have a common orientation. The cleavage totally pervades the rock and consequently is termed a pervasive cleavage. The rock-cleavage is thus a product of the rock-fabric or texture. It differs from bedding fissility, described above, because the orientation of the schistosity is controlled by the directions of the tectonic forces and movements. Like fracture cleavage, it is generally oblique to the stratification. It is produced by a combination of mechanical rotation and recrystallisation of the rock components. Any new minerals developed during this process have the compositions and assume the orientations which are most closely in equilibrium with the physico-chemical conditions existing in the rock at the time.

The development of both fracture- and flow-cleavage in rocks is a function of first, the type of rock; secondly, the intensity of the tectonic deformation; and thirdly the depth of burial and the metamorphic environment.

In rocks having different lithologies, the stronger the rock, the less likely it is to be cleaved. Flow cleavage tends to develop more readily in the more easily deformed rocks which are fine-grained, such as shales or tuffs. These not only deform without much difficulty, but also recrystallise more easily than tougher strata like sandstones or lavas. In a group of mixed strata one commonly finds that the more rigid beds are cut by a fracture cleavage while the less rigid are changed to slate. This results in the group as a whole showing differential cleavage, in which the cleavage planes are refracted as they pass from one bed to another of different composition. Isolated occurrences of cleavage, in formations which are dominantly uncleaved, may arise in beds which are unusually weak. Fourmarier (1948) has found that a single coal seam lying in rocks which show little sign of serious deformation, can develop a cleavage; Slater (1927) also noted a strain-slip type of cleavage in some of the contorted glacial clays of East Anglia.

The intensity of the tectonic deformation has naturally a marked effect on the development of cleavage. Thus, in a series of strata, should the movements or stresses be concentrated in one or more particular zones in the rocks, or on certain horizons, one finds that there the cleavage is more strongly developed than elsewhere: this accounts for the common formation of a local cleavage or schistosity in the vicinity of faults or in zones of thrusting (Fourmarier 1948, Wilson 1951, Fig. 13b, Bloomer & Werner 1955, Aderca 1960).

The importance of the load on the formation of rock cleavage has been demonstrated by Fourmarier in numerous publications (1923, 1932, 1952). He has clearly shown that in rocks of similar type, in the same tectonic region, cleavage and schistosity do not begin to develop until a certain minimum depth of burial has been exceeded. This accounts for the fact that though a group of rocks may be strongly folded, they may still show little or no sign of the development of a cleavage, for example, in the Jura. The depth at which cleavage begins to appear depending on local conditions, lies somewhere about 5000 to 7000 m, and it forms an **upper front of cleavage schistosity***. The first indication of this front is the wide-spread development of fracture cleavage in the less resistant beds, which continues downwards in depth until, at a lower front, it is

---

* Kienow (1942) has suggested a depth of 3000 to 6000 m for the zone of fracture cleavage and schistosity in slates (*Transversalschieferung*), and over 6000 m for the crystalline schists and gneisses developed with the aid of recrystallisation processes. In the Pyrenees, crystalline schists appear to have developed under a load of not more than 3000 m (Zwart 1954).

replaced by slaty cleavage. Below this slaty cleavage is the zone of crystalline schists. The positions of the various fronts do not necessarily conform to the stratification of the rocks, but they reflect the depth of burial at the time of the movements which were responsible for the cleavage structures. For example, Fourmarier has noted the manner in which the rocks in certain Alpine nappes may be strongly schistose, though rocks of similar age and composition in the overlying nappes are free from cleavage (Fourmarier 1953). He has also pointed out (1949a, p. 659) that it is quite possible for rocks which have been rendered schistose in an early phase of a tectonic episode to be thrust over uncleavaged rocks during a later phase of the same episode.

The effects of metamorphism, migmatisation, and igneous intrusion can locally modify the levels of the various fronts of cleavage or schistosity (Fourmarier 1959). In the neighbourhood of large intrusive bodies, where the physical state of the country rocks has been influenced by heat or by emanations, the levels of the fronts tend to rise. One may therefore find a rapid and, at first sight, an apparently anomalous increase in the intensity of the schistosity and deformation as one enters the metamorphic aureole of an intrusion, or approaches a migmatite complex. It is, however, not essential that the deformation of the country rocks which surround a large igneous body, nor that the schistosity which occurs in them, should be related to the regional tectonics. Such structures have, in many localities, been formed directly by pressure or shearing stresses exerted by the intrusion itself during its emplacement. Some of these deformed and schistose zones, in the country rocks around big batholiths, have widths of up to 3 km and the intrusion itself may also be deformed and foliated for similar distances inwards from its edge (Balk 1937, p. 67, Read & Pitcher 1960).

To the geologist in the field, the great significance of well developed fracture cleavage or schistosity in folded beds is, as the 'Old Men' fully realised, that the surfaces of the secondary structure lie subparallel to the axial-planes of the folding — provided that the structure and the folding were generated during the same tectonic episode*. If the folds are not plunging then the strike of such axial-plane cleavage is thus parallel to the general direction of the fold hinges, and so, no matter how the beds themselves are plicated, one still has a feature in the rocks which will act

---

* 'Where the cleavage dips at a high angle, the contortions also have their axes similarly inclined; whereas, when it is nearly horizontal, so also are their axes' (Sorby 1853, p. 139). Sorby's use of the term **axis** where we would use **axial-plane** makes this statement appear somewhat confusing at first sight.

as a guide to the trend of the large-scale structure (Darwin 1846). A swing in the direction of the cleavage commonly denotes a corresponding swing in the direction of the folds, an observation which was used in the construction of a tectonic map of North Wales by Shackleton (1953). Similarly, changes in the dip of the schistosity across an area normally indicate a similar change in the asymmetry of the folding unless the fan-shaped arrangement of the planes is strongly marked. Even before the true style of the folding can be recognised by the mapping of the beds, one can, by recording the strike and dip of the schistosity over a relatively small area, arrive at close approximations to the asymmetry, direction, and *régard* of the major structures in the neighbourhood. Abrupt changes in the orientation of schistosity very commonly indicate that the rocks of an area have been deformed by some kind of later movement. For these reasons alone it is worth mapping schistosity with the same care and precision that one normally pays to stratification.

In regions where folding and schistosity or cleavage have resulted from the same movements, that is, provided the cleavage is not superimposed on an earlier fold system, one can, by observing the relationships between the cleavage and the stratification, deduce not only the geometrical form of the folds but also the stratigraphical succession. Thus, in an area of overturned folds the dip of the normal limbs is less steep than the dip of the fold axial plane; and the dip of the inverted limbs is steeper than that of the axial planes (Figs 6.1a, 6.3). Consequently, if the schistosity or cleavage is parallel to the axial planes of the folds, we can say that: *Where the schistosity and the beds dip in the same direction, if the schistosity is steeper than the stratification, the succession is in the correct order; if the schistosity dips less steeply than the bedding, then the beds are probably inverted.* This rule was illustrated by an example observed by Fourmarier, (1932) from the neighbourhood of Ventura, California. At the point where he made his observations the beds dipped at 80° towards the north, and the schistosity dipped at 60° in the same sense. These beds were inverted and formed the northern flank of a syncline overturned towards the south. Other examples of the relationship between stratification and cleavage on the overturned limbs of folds are shown in Figures 6.1a and 6.3 and the left-hand side of Figure 7.7, the Loch Alsh syncline, Scotland; and Figure 6.2, Anglesey. Even if the schistosity tends to form a fan across a fold, as it often does, these relationships are unchanged.

These general rules, which were enunciated by Leith, do not hold for those areas where the beds were not in their correct order of succession before the folding, with which the axial-plane cleavage is associated, took place; nor can they be applied blindly to recumbent folds with

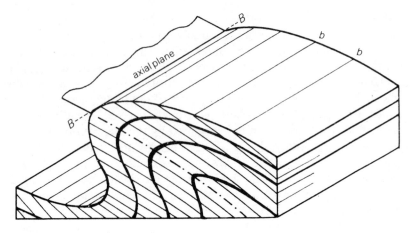

Figure 6.3 The relationship between schistosity parallel to the axial plane, and bedding in an overturned fold.

plunging brows. But they can be used to elucidate the geometry, if not the stratigraphy, of antiforms or synforms in unfossiliferous rocks.

The relationship of axial-plane cleavage or schistosity to stratification in a recumbent fold with a *tête plongeant* is shown in Figure 6.4a. In the root-zone, where the fold is rising towards its crest, the structure faces upwards, and the schistosity–stratification relationship is similar to that in a normal fold. As the fold drops from its crest towards the hinge, the relationship becomes reversed: in the normal or upper limb, the schistosity dips less steeply than the stratification; while in the inverted or middle limb of the fold, the schistosity dips more steeply than the bedding.

An example of such an anomalous schistosity–stratification relationship is shown in Figure 6.4b and c. The beds illustrated are schistose Upper Jurassic limestones which form part of the upper limb of the Morcles Nappe of the Helvetic Alps. The locality is situated on the side of the road to the Gemmi Pass, near the south end of the Dauben See. The beds dip about 30° to the north-west, and are in their correct order of succession; nevertheless, the schistosity, as seen in section, is nearly horizontal because the exposure lies between the crest of the Nappe, where it arches over the south-west end of the Aar Massif, and the fold-hinge which lies further to the north-west and at a lower level. The general structure is shown in Figure 6.4c (see also Fourmarier 1953b).

Shackleton drew attention to the criteria by which, given sedimentary structures such as graded bedding, together with axial-plane cleavage, one can distinguish whether the major structure faces upwards or downwards (Shackleton 1957). He found that if, when one considers the

Rock cleavage and schistosity

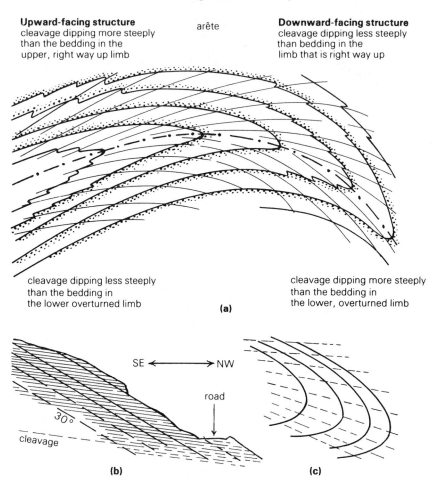

Figure 6.4 (a) The relationships of stratigraphy and schistosity in a recumbent fold with a plunging brow. (b), (c) The relationship between stratification and schistosity in the upper limb of the Morcles Nappe, near the Gemmi Pass.

limbs of a fold, the schistosity–stratification relationship accords with the sequence shown by the sedimentary structures, the fold faces upwards (Fig. 6.5a). If the sedimentary structures and the schistosity–stratification relationships are non-accordant, then the folded structure faces downwards (Fig. 6.5b).

The line of intersection of the planes of axial-plane cleavage or schistosity and of the stratification has been shown by Leith (1923, p. 127) to be parallel to the plunge of the local fold axes (Fig. 6.6). In a small exposure, the schistosity and the stratification together form a

46 *Small-scale geological structures*

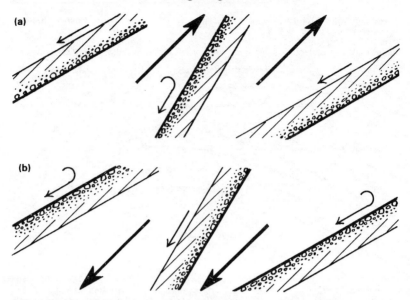

Figure 6.5 Upward- and downward-facing fold structures. (a) Upward-facing folding: the cleavage and sedimentation structures are in accord. (b) Downward-facing folding: the cleavage and the original sedimentation structures do not accord (after Shackleton 1957).

monoclinic structure in which the b-axis will be parallel to the B-axis of the major structure. This relationship has widespread application, because it is not uncommon for the actual fold-hinges to be masked by overburden, and if numerous observations of this linear structure are

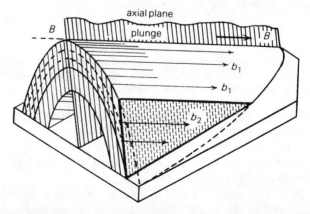

Figure 6.6 The relationship between the axial plunge of a fold (B) and the lines of intersection between bedding and cleavage: b_1 is the trace of the cleavage on the bedding, and b_2 the trace of bedding on the cleavage.

made it may be possible to recognise changes in the direction and plunge of the fold-axes which otherwise could not be demonstrated.

Though the relationships between cleavage or schistosity and folding discussed above hold good in very many areas, there are exceptions. These are most common in regions where shear-folding — *Scherfaltung* or *Gleitbretterfaltung* — has been predominant. Such folds result from differential slip of the rocks on numerous closely-spaced planes of movement. Here, not only may the amount of slip vary from place to place along the strike of the slip-planes, but the beds themselves may have been deformed to a greater or lesser extent during early stages of the tectonic movement. Under these circumstances the resulting structure is commonly triclinic, and not monoclinic and homoaxial. In consequence even though the cleavage is parallel to the axial plane of the visible fold, the *a*-direction of movement on the cleavage need not necessarily be at right angles to the *B*-direction of the fold itself. The problem is one of superposed structures and will be considered later. Discussions of the problem and its solution have been given in Ramsay (1960).

7 Fracture cleavage and strain-slip cleavage

Fracture cleavage and strain-slip cleavage are both non-pervasive cleavages, i.e. the individual cleavage planes are separated by uncleaved slices of rock termed microlithons. Nevertheless, they appear differently in the field and beneath the microscope; the microlithons associated with fracture cleavage are not folded, those associated with strain-slip cleavage are. Both are included in Fourmarier's definition quoted on p. 36; but, as I suggested there, fracture cleavage should be retained for the closely-spaced planes, which may be irregular and discontinuous, or which may be net or clean-cut occurring in homogeneous rocks; and strain-slip cleavage should be used for the secondary or false-cleavage developed in those rocks which already possess a primary schistosity. Strain-slip cleavage thus corresponds to Born's *Runzelclivage*, Knill's crenulation cleavage, and Dale's slip cleavage.

Fracture cleavage is normally confined to rocks which are virtually unmetamorphosed, or which show a low-grade of metamorphism only. Though it is generally associated with the folding of strata, it may also develop in the vicinity of fault- or thrust-zones.

In its normal form of development, in stratified rocks, or in a zone of dislocation, the cleavage appears as a series of parallel plane fractures. In massive, brittle rocks these fractures may be so widely spaced as to merit the term 'close jointing'; but in beds which are relatively weak the cleavage may be so fine that up to 150 fracture planes can be contained in one centimetre. The cleavage is formed mechanically, and the rock-slices between the planes of fracture are not necessarily metamorphosed during its formation.

Fracture cleavage associated with folding generally forms an acute angle θ with the bedding. However, the cleavage is at right angles to the stratification, at fold-hinges or in tough or competent beds.

The angle θ between the cleavage and bedding planes commonly reflects the lithology of the bed which is cleaved, Figure 7.1. In incompetent beds it is small, but in more rigid beds it may approach 90°.

Fracture cleavage and strain-slip cleavage 49

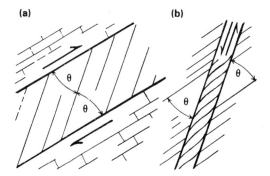

Figure 7.1 (a) Cleavage and bedding: the general sense of the slip along the stratification is indicated by the acute angle ϕ. (b) The sense of slip on a normal fault as shown by the relationship of the cleavage to the plane of movement.

This leads to **refracted cleavage**, in which the cleavage planes are abruptly deflected as they pass from one bed to another (Fig. 7.2a). This is commonly seen in regions where such rocks as interbedded grits and shales have been folded together (Fig. 7.2a). Changes in the lithology within a single bed, such as one finds in graded graywacke strata, lead to the formation of curved cleavage slices. In the coarse arenaceous lower part of the beds, the cleavage forms at a large angle with the stratification; but in the finer-grained, more argillaceous upper portion of the beds, it bends or curves over so as to meet the overlying bedding plane at a relatively small acute angle (Fig. 7.2b).

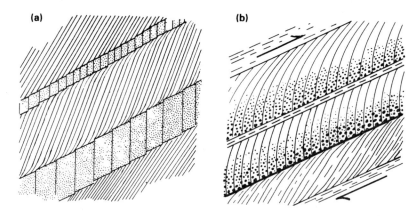

Figure 7.2 (a) Refracted cleavage in interbedded rigid and plastic strata. (b) The curvature of cleavage developed in a graded bed.

In some argillaceous beds which have been folded, either the bedding fissility, or the stresses developed by the bending of the beds have caused the rocks to fracture in two directions: one may be parallel to the stratification, the other forms a normal transverse fracture cleavage. The combination of these two surfaces of rupture breaks the rock into long, narrow strips which have rectangular or rhombic cross sections. The structure is known in English, as **pencil cleavage** (*Griffelschiefer*, Born 1929, p. 350); and the lengths of the 'pencils' are parallel to the local fold axes.* Because the cleavage refracts as it passes from one bed to another it is not truly axial-plane cleavage except in the hinge region of the fold. However, in the less competent strata, the angle between the cleavage and planes of stratification is often very acute, so much so that it may form parallel to the axial plane and a true axial-plane cleavage may develop. More commonly, however, the fracture cleavage forms a fan which is symmetrically oriented on either side of the axial plane (Fig. 7.3).

With the growth of the fold after the initial fractures have been formed, the thin slices (*feuillets*) may move relative to each other in response to the further slip between the beds. On the flanks of the fold each cleavage plane may act as a minute normal fault, and one may observe the bedding planes corrugated by innumerable fault-scarplets parallel to b . . . 'so truly reflecting the form of a series of steps' (Laugel 1855). This movement, a combination of slip and rotation, results in the bed becoming thinner normal to the stratification, and in its extension in the general direction of movements (Fig. 7.4a and b). At the same time this rotation of the cleavage slices brings them more and more closely towards parallelism with the axial plane of the fold, until a true axial-plane cleavage may be produced. In exceptional cases the cleavage slices may even be rotated so far that they converge toward the fold-hinge, as seen in the thin shale band in Figure 7.3. Again, the thin slices in the weaker beds will tend to be more affected than the thicker slices in the more rigid strata; thus further emphasising the refraction of the cleavage between beds of different lithology. As pointed out by Lovering (1928), the rotation of the slices can only take place if the upper and lower ends of the fracture cleavage planes overlap (Fig. 7.4a and b). If they are more widely spaced, the rotation of the rhomb-shaped slice would result in the thickening of the bed against the weight of the overlying strata, which would be contrary to the principle of the expenditure of minimum energy, (Fig. 7.4c and d). Not uncommonly the cleavage slices are deformed after or during their formation. Their terminations are then

---

* Pencil cleavage *in slates* is usually at right angles to this direction, see page 63.

Fracture cleavage and strain-slip cleavage

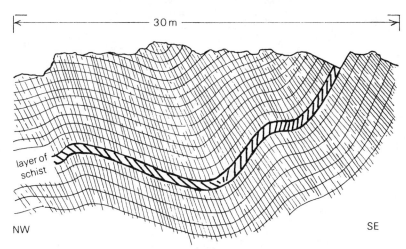

Figure 7.3 Cleavage fan in a syncline of Torridonian arkoses, Isle of Islay, Scotland.

drawn out and sigmoidally curved in the direction of movement (Laugel 1855, Leith 1923, Figs 57 & 58, Wilson 1951, Fig. 7) (Fig. 7.5).

Other workers considered that fracture cleavage was formed by shear failure of the rock. However, the mechanism of formation of fracture cleavage is now considered to be similar to that of flow- or slaty-cleavage, the factor governing which of the two develops being primarily one of lithology. Fracture cleavage forms in more arenaceous rocks, slaty cleavage forms in more argillaceous rocks. The way in which a transition from one type of cleavage to the other can occur in a graded bed is shown in Figure 7.2b. The cleavage planes are thought to coincide with the flattening plane (AB) of the finite strain ellipsoid. This has been convincingly demonstrated by Dieterich and Carter (1969) who use the technique of finite element analysis to determine the orientation of the AB plane of the finite strain ellipsoid at various stages in the growth of a fold and compare these orientations with the orientation of cleavage around natural folds.

Fracture cleavage is often found in association with folding. However, it is sometimes found in association with thrusts or other faults. A good example of the development and subsequent deformation of fracture cleavage, in a region where shearing has been the dominant active force, can be seen to the SSW of the Schwarenbach Hotel on the footpath towards Rote Kumme, in the Gemmi Pass area of Switzerland*. Here the

---

* I would like to take this opportunity of thanking Professor Eugène Wegmann who kindly introduced me to this area.

Small-scale geological structures

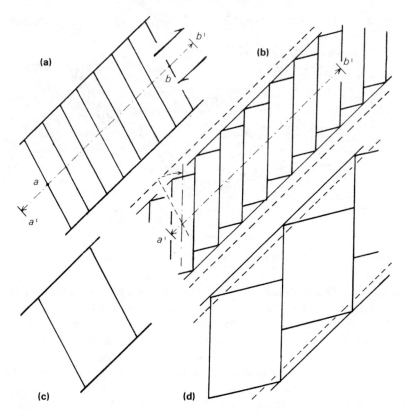

Figure 7.4 The spacing of cleavage planes. (a), (b) The thinning and extension of a cleaved bed from (ab) to (a^1b^1) by the rotation of the fracture cleavage slices. (c), (d) How the rotation of thick cleavage slices is prevented because it would result in the thickening of the bed against the weight of the overlying rocks.

Hauterivian of the Nappe de Morcles consists of siliceous limestones and calc-shales, which dip about 35° to the WNW. Above these lies the Nummulitic, which forms the envelope of the nappe, and which, in turn, is overlain by the Nappe de Diablerets-Gellihorn. The direction of movement of the upper strata relative to those below was towards the WNW that is, down the dip. Two exposures are significant; in the first (Fig. 7.6a), a bed of calc-shale lying between beds of limestone is traversed by a strong fracture cleavage which is perpendicular to the stratification.

In the second exposure (Fig. 7.6b and c), which is situated a short distance to the south-west of the one described above, the geological setting is the same; but here, the cleavage has been dragged over by movement of the overlying nappe. No evidence was observed to suggest that the curvature of the cleavage might have resulted from graded-bedding in the calc-shale.

The formation of fracture cleavage is not only confined to folding. It may develop on fault-zones, and it is common in the vicinity of big thrusts, where the cleavage may traverse the rocks above and below the plane of movement for considerable distances. The origin of such non-penetrative cleavage is not always clear. It may be parallel to the fault-plane, thus forming a **sheeted fault zone** or it may lie at an acute angle to it, as shown in Fig. 7.1b.

The mechanism of formation of this type of cleavage is not clear. It may be formed by shear failure of the rock adjacent to major planes of movement; many examples of significant displacements on these cleavage planes have been recorded. The development of local planar fabrics (cleavage) in association with shear zones has been discussed by Ramsay and Graham (1970). Here the cleavage forms originally at 45° to the direction of movement in the shear zone and like the fracture cleavage formed in association with folding, can be related to the flattening plane (AB) of the finite strain ellipsoid.

Becker (1882) demonstrated the effects on the country rocks of closely spaced shear-zones which were parallel to, and which graded into faults. He showed how differential slip along such planes offset the strata which they cut, and that folded beds could be produced by this mechanism. The structures so formed are closely related to Schmidt's *Gleitbretterfalten*. Becker found by mathematical analysis and experiment, that the displacements of the beds across a fault-zone which was made up, in this manner, of a large number of parallel planes of slip, closely approximated to a logarithmic curve. The form of the curve thus developed in bedded rocks depended on the angular relationship between the plane of the fault and the stratification. In these structures the cleavage and the movements on it were responsible for the apparent folding, but the cleavage itself was a subsidiary of the fault plane: it bore no direct relationship to the orientation of beds which it cut. Consequently, the stratigraphical relationship between the cleavage and the bedding does not hold good.*

On some dislocations, an oblique cleavage may be developed. This cleavage may be confined to the shattered rock within the fault-zone itself, or it may appear in the rocks above or below the plane of movement. Examples of the former have been illustrated in Nevin (1936) and Wilson (1951, Fig. 7, p. 409). Numerous examples, seen in the

---

* An example of schistosity parallel to a thrust is illustrated in Kautsky (1953).

Figure 7.5 Dislocation cleavage in the root-zone of the Helvetic Nappes, Saillon, Valais, Switzerland.

'*bassin houillier meridional de la Belgique*' and locally associated with boudinage, have been described by Aderca (1960). He has referred to the structure as *schistosité de dislocation*.

Where the cleavage occurs in the rocks outside the zone of dislocation (Sheldon 1928) it is not necessarily connected with the regional folding, and may even cut cleanly across both the limbs of folds lying below (or above) a thrust. The local relationship between cleavage and stratification, as seen in isolated exposures, could thus be very misleading, if the origin of the structures were not fully appreciated.

An example of cleavage associated with thrusting has been mapped by Kanungo (1956) in the Kyle of Lochalsh area of Ross-shire, NW Scotland. This area lies within the Moine Thrust Zone, and in it are two subsidiary thrusts which carry Lewisian Gneiss (Fig. 7.7) over a

Fracture cleavage and strain-slip cleavage

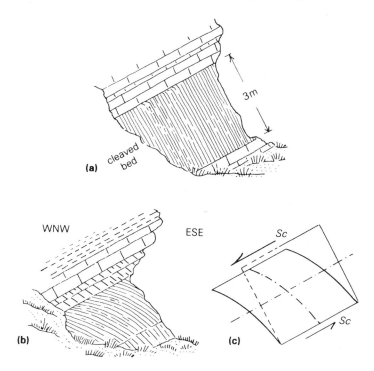

Figure 7.6 The development, and subsequent curvature of fracture cleavage, near Schwarenbach, Switzerland.

recumbent syncline of Torridonian Sandstone. This fold is generally referred to as the Loch Alsh Syncline; it closes towards the east, so that the gently inclined and inverted limb also dips easterly. The hinge of the fold which trends roughly NNE can be seen in several places; and in addition current bedding and slump structures in the Torridonian beds clearly demonstrate that the strata, which lie above the core of the fold, are upside-down. A normal fracture cleavage appears in the rocks near the hinge, and its relationship to the stratification is exactly what one would expect under the circumstances; remote from the core of the fold, this cleavage disappears. But, as one proceeds eastwards, a new cleavage makes its appearance, near the meridian of Loch Iain Oig, in the more argillaceous beds of the Diabaig Group which form the (inverted) Lower Torridonian. It dips easterly at a steeper angle than the bedding, and at first sight suggests that the beds there are in their correct stratigraphical sequence, and that the fold-hinge which lies to the west should be that of an anticline closing to the west. This cleavage, which becomes more prominent as one goes eastwards, owes its formation to the movement on

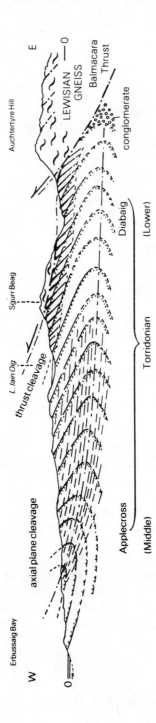

Figure 7.7 Sketch section of the structure near Kyle of Lochalsh, showing the normal cleavage developed in the core of the Loch Alsh Syncline, and thrust cleavage formed below the Sgurr Beag Thrust (after Kanungo 1956).

the gently dipping Sgurr Beag Thrust, which though now eroded, once lay about 100–150 m above the exposures where the cleavage can first be seen.

The development of a coarse fracture cleavage which is the direct result of thrusting has also been described by Bloomer and Werner (1955). These authors observed that the structure was 'essentially a very close-spaced imbrication of thrust-faults and it cannot be used to determine stratigraphic succession'. The cleavage strikes parallel to the strike of the thrusts, and dips at angles which vary from 70° to 20°. In greywackes and tuffs, the cleavage planes are spaced at 1 mm apart, in quartzites and coarse-grained rocks the fractures may be separated by a metre or more. The displacements on the fracture planes also vary widely, from less than 2.5 cm to several metres. Bloomer and Werner have suggested the term **thrust cleavage**, (*clivage de charriage*), for this structure. The well developed cleavage which can be seen at Saillon, in the root zone of the Helvetic Nappes, is probably of this type (Fig. 7.5).

The trace of the cleavage on the plane of dislocation declares the *b*-direction of the structure, and it lies at right angles to the direction of movement. The sense of the movement is shown by the acute angle which the cleavage makes with the fault plane. The plunge of the *b*-direction can commonly be observed; elsewhere it must be computed from the dip and strike of the cleavage and of the surface of movement.

Strain-slip cleavage, like fracture cleavage, is a non-pervasive cleavage. It differs from fracture cleavage in that the slices between the cleavage planes (the microlithons) are composed of rock which is fissile or schistose; these pre-existing surfaces (S_1) are now intensely puckered. The spacing between the cleavage planes may vary widely. In some localities, like that shown in Figure 6.1b the cleavage occurs in association with zig-zag folds. Heim (1921) illustrated examples from the Punteglias valley, Switzerland, where the cleavage planes lie about 1 cm from each other. Balk (in Balk & Barth 1936, p. 714) interpreted the strain-slip cleavage as shear planes and recorded that the 'shear planes' were spaced at intervals of 7–13 mm, although some are as much as 26 mm apart. Fifty-one planes were measured in 52 cm of schist. In other regions the spacing is so small that it can only be observed beneath a microscope.

The majority of observers consider that the structure has developed along the planes of weakness in the rock formed by the plications of pre-existing surfaces. Thus Cloos and Heitanen (1941, pp. 158–9) found that a cleavage developed where small-scale folds graded into crinkles, of which the order of magnitude approached the size of the mica flakes in the original schist. Elsewhere, however, one may observe examples in

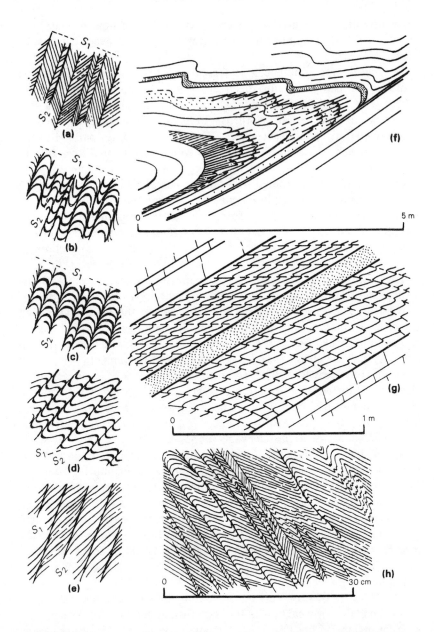

Figure 7.8 (a)–(e) Varieties of small-scale folding with which strain-slip cleavage is associated and examples of field occurrences. (f), (g) The cleavage occurs in pelitic, and semipelitic crystalline schists, Moine Series, Ross of Mull, Argyllshire, Scotland. (h) Strain-slip cleavage in an actinolite schist, A'Mhoine, Sutherland, Scotland.

which the slip on the cleavage surfaces appears to have dragged the early schistosity into parallelism with the cleavage planes. It seems unwise to state categorically which is cause and which is effect (Fig. 7.8).

The planes of strain-slip cleavage are commonly loci of recrystallisation. The platy minerals of the early schistosity (S_1) are bent into parallelism with the later planes of cleavage (S_2), which, in turn, may act as privileged paths for mineralising solutions. There is a tendency for the preferential growth of those minerals which lay closest to the cleavage surfaces, and for the production of new minerals along those planes. The process of recrystallisation and growth of new minerals will tend to destroy the early schistosity and there may be a complete gradation from strain-slip cleavage cutting across an early schistosity or stratification, to a crystalline schist in which the earlier structures are obliterated.

The small-scale folds with which strain-slip cleavage is associated show a variety of forms (Fig. 7.8a–e).* Commonly the cleavage coincides with the axial planes of chevron folds (Figs 6.1b, 7.8a); and its development follows the manner so clearly described by de Sitter (1956, p. 94 *et seq.*, Figs 65, 66), though he refers to the structure as 'fracture cleavage'. A good example is shown in Hills (1953, Fig. 66A). The planes of weakness along which the cleavage has developed coincide with the abrupt bending of the rock at the hinges of the micro-folds.

In some rocks, tight sinusoidal folds (Fig. 7.8b) rather than chevron folds, have been produced, and the planes of weakness follow the fold-limbs, where the original schistosity (S_1) has been bent to form new, nearly continuous surfaces of parting (S_2). Any slip that occurs will then take place along the fold limbs, and the slices between the cleavage planes will contain nearly complete micro-antiforms or synforms. Examples are shown in Balk and Barth (1936, Figs 15–18), Heim (1878, Atlas, Pl. XV, Fig. 11, and 1921, II, Pl. II, a and b), Kienow (1942, Pl. IV) and Gonzalez-Bonorino (1960). Closely akin to this sinusoidal structure is that shown in Figure 7.8c and f, in which one group of hinges has been eliminated, and the plications form a series of micro-coulisse folds. This style of cleavage is commonly observed in the hinge zones of major folds, where it occurs as an axial-plane cleavage. Under these circumstances, it can be used to determine the geometry of the major fold, and perhaps even the stratigraphical succession.

Strain-slip cleavage may also develop in association with small-scale asymmetrical folds. Here the planes of weakness in the rocks are formed

---

* Some examples of strain-slip cleavage are very similar in appearance to forms of *microplissement* and *plis miniscules* described by Professor Fourmarier: compare Figure 7.8 with illustrations in Fourmarier (1951, 1952a). The genetic differences between these structures is discussed on p. 66–7.

by the attenuation of the steep or inverted limbs of the micro-folds (Fig. 7.8d, g and h). The structure suggests a yielding of the rock by shear as a result of compression or torsion, analogous to thrusting; the asymmetry of the micro-folds declares the sense of movement on the cleavage planes.

The manner in which the early schistosity has been bent may also suggest that the deformation has resulted in a faulting movement throughout the rock mass (Fig. 7.8e). Here the schistosity (S_1) is not folded in the strict sense, but is curved sigmoidally; this variety was well illustrated by Born (1929, Pls IV, V), and is also shown in Harker (1932, Fig. 68).

8 Flow cleavage, schistosity and lineation

The formation of flow cleavage or schistosity results from the recrystallisation of the rocks while they are suffering internal deformation or movement, caused by the application of external stresses. Indeed I doubt whether a true schistosity could be well developed in rocks where a relief movement was inhibited; at the most it would then become little more than a tectonic compaction, though this might be accentuated by metamorphism. In nearly every case where schistosity is strong, the worker can generally find evidence that the rocks have not only been powerfully compressed, but they have also been deformed and drawn out in the direction of easiest relief. Sharpe (1947) and Sorby (1853) both emphasised the importance of this elongation or stretching in the production of slaty cleavage; but in many textbooks this factor is ignored, and reference is only made to the fact that the rocks have been subjected to pressure. The importance of, and the evidence for such stretching have been summarised by Fourmarier (1949b).

Flow cleavage or schistosity which is associated with folding though not necessarily in high-grade metamorphic terrains is commonly parallel or subparallel to the axial planes of the folds. In consequence it can be used in the elucidation of the local structure, and in the determination of the stratigraphic succession of the folded beds, as outlined in Chapter 6 pp. 42 *et seq.*

The development of schistosity in rocks has for many years been a subject of discussion among geologists. Harker's Report to the British Association for the Advancement of Science in 1886 includes an admirable account of the formation of slaty cleavage, and it summarises the results of the early investigators of this structure. Other contributions to the problem include Van Hise (1896) on the *Deformation of rocks*; Leith's bulletin on *Rock cleavage* (1905), and passages in his textbook on *Structural Geology* (1923); Behre (1933), *Slate in Pennsylvania*; Mead (1940); Swanson (1941) on *Flow cleavage . . .*; E. Cloos (1937), and his paper on *Oolite deformation in the South Mountain Fold, Maryland*

(1947); Goguel (1945); Wilson (1946); Colette (1958) and Siddans (1972). Many papers discussing the various aspects of schistosity have been published by Professor Fourmarier and these have been summarised by Baer (1956). All modern textbooks describe the formation of schistosity, to a greater or lesser extent; and serious-minded students will find detailed discussions on the problem of deformed rocks, and on the developments of schistose structures in works on structural petrology or *Gefugekunde*: Sander (1930, 1948, 1950); Schmidt (1932); Knopf and Ingerson (1938); Coles Phillips (1937); Turner and Verhoogen (1951, Chs 20, 21); and Voll (1960); all of which contain extensive bibliographies on the subject. An excellent review on slaty (flow) cleavage is given by Siddans (1972).

The most simple variety of flow cleavage is that seen in roofing slates, where the rock splits cleanly, parallel to the schistosity, in smooth slabs. Such rocks are the product of low-grade metamorphism, and lie near the upper front of schistosity. At greater depths, or under more severe metamorphic conditions, the rocks develop a more coarsely crystalline texture. They split less regularly, the cleavage surfaces have a satin-like sheen, and are commonly undulating or may be finely corrugated: these are phyllites, or sericite- or chlorite-schists; and they in turn grade into the realm of the crystalline schists. Nearly all the crystalline schists possess one good, easy direction of splitting, but the schistose surfaces are almost always crinkled by at least one variety of linear structure. Some of the coarser varieties of crystalline schists are foliated, with their schistosity parallel to the original stratification; consequently it may be very difficult to differentiate between schistosity and bedding, even in the hinges of folds. Nevertheless, a careful search of apparently coincident structures may permit one to distinguish between the two, even when the angle between them is very small.

Nearly all good roofing slates possess in addition to their main planes of schistosity, a second direction in which they can be broken with relative ease. This is known as the **grain**. In some rocks it shows as a striation on the surfaces of schistosity; in others it is barely discernible to the naked eye. Beneath the microscope one sees that, parallel to this direction, grains and flakes of sericite and chlorite have a marked elongation, quartz is drawn out into narrow spindle-shaped rods, and rutile needles have a similar orientation. The grain is thus determined by the parallel elongation of the crystals in much the same way as the schistosity results from the parallelism of their cleavages. The relationship of this direction to the movement which the rocks had suffered was clearly demonstrated by Jannetaz (1884).

Jannetaz observed that not only was the grain parallel to the direction

Flow cleavage, schistosity and lineation 63

of maximum elongation of fossils in a slate, but also that it was parallel to the maximum thermal conductivity of the rock. He then proceeded to investigate the phenomenon experimentally by compressing a soft clay in a pressure-box, the top of which was left open, so as to give an easy relief normal to the direction of applied pressure. The clay was thus compressed in one direction, and squeezed out of the box in the other. Jannetaz then found that the maximum thermal conductivity of the deformed clay was parallel to the direction of easiest relief, and was least in the direction of greatest compression. The original isotropic clay had thus assumed an anisotropy similar in every way to that found in slates, but without the medium suffering any recognisable metamorphism other than simple deformation. The fundamental molecular fabric of the material was, in this way, decided by the deformation alone. With recrystallisation under metamorphic conditions, the rock-fabric would naturally become coarser, schistosity could be seen, and the grain could be recognised beneath the microscope or in the hand-specimen.

The trace of the grain on the surface of schistosity in slates is therefore a linear structure which declares the direction of maximum extension in the rock and lies at right angles to the b-tectonic axis (Fig. 6.3). In terms of the strain ellipsoid, the schistosity corresponds to the AB-plane (Fig. 8.1), and the lineation formed by the grain on the plane of schistosity is parallel to the AA-axis of the ellipsoid. Where the grain is well developed its intersection with the schistosity may result in a pencil-cleavage parallel

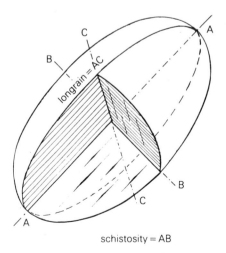

Figure 8.1 The relationship between flow-cleavage schistosity, longrain, and the strain ellipsoid.

to the maximum extension direction. Jannetaz mentions that locally in the Pyrenees it is so strong that the slates can be split in two directions to form plant supports (*échalas*).

The megascopic structures which indicate the maximum extension direction are well known (Behre 1933, Cloos 1946). Deformed fossils have been described and measured for over 100 years, and Heim's famous illustrations of belemnites have been reproduced in publications all over the world (Sharpe 1846, Sorby 1855, Heim 1878, Jannetaz 1884, Goguel 1952). Pebbles drawn out in the direction of maximum extension were recognised by Ramsay (1881) in North Wales, but had previously been discussed by Hitchcock and Hager (1861) in the United States. More modern descriptions include several examples from the Caledonides, Strand (1945) and Flinn (1956, 1961) who have made analytical studies of the phenomenon. Deformed lava pillows, amygdules and agglomerate fragments all drawn out and elongated in the maximum extension direction were described by Wilson (1951, p. 404); Cloos and Heitanen (1941, pp. 83–4) observed the elongation of volcanic spots which were '. . . paper-thin up to 1 inch wide (2.5 cm) and a foot long (30 cm), thus forming lenticular bodies whose longest axes are strictly parallel . . . The lineation (elongation) is within the cleavage plane and thus independent of the bedding . . . Its trend is roughly perpendicular to the axes . . . of the folds' (quoted also in Wilson 1946). The most valuable work on this general topic is undoubtedly E. Cloos' study of deformed oolites in the South Mountain Fold of Maryland (Cloos 1947b).

The significance of deformation, extension in one direction and compression in another, has rarely been taken into consideration when estimates of the thicknesses of strata have been made in regions where folding has been accompanied by the development of schistosity. Green (1917) calculated the extension of the 'eyes' in the 'birds' eye slates' of the volcanic rocks in the English Lake District, after having assumed that the ellipses he observed were originally spherical. From these measurements he concluded that one of the local tuff horizons, which measured 215 m thick, had an original thickness of 155 m. Cloos (1942, 1943) making use of a large number of measurements of deformed oolites in the Appalachians found that several groups of beds had been greatly thickened near the hinges during folding. He even suggested that in some places the strata would need to be reduced to 50 or 60 per cent of their measured thicknesses. In many regions, however, the absence of measurable evidence, from which the deformation might be computed, leaves one no alternative but to make a guess at any changes in thickness which the beds may have undergone.

Flow cleavage, schistosity and lineation 65

With the development of schistosity in a rock-series which is being folded, several new factors may make their appearance. For instance, the recrystallisation of the rock will change its physical properties, so that the originally soft, easily deformed material will tend to become more resistant and more rigid, due to 'work-hardening'. Folding by flexure may then cease, but if the pressures are still active, the schistose mass as a whole may fracture by shear, with the formation of a 'shear-cleavage' or the development of normal kink bands (Ch. 7), which will appear in the rocks as a strain-slip cleavage, cutting across the schistosity of the slate, and perhaps ruining large parts of it for economic exploitation.

In the early stage of folding, the most important planar fabric is often bedding. This fabric plays an important role in the folding. Initially, the folds develop by flexural slip, a process which involves the sliding of the beds over each other (and which produces folds with a parallel or concentric profile geometry). However, as folding continues an axial-plane fabric develops and the bedding fabric becomes less pronounced. Consequently, folding by flexural slip becomes less important and eventually ceases. The folds may, however, continue to amplify under the continued action of the compressive stress, by the process of homogeneous flattening during which the axial-plane fabric becomes more and more pronounced and the fold geometry changes gradually from that of a parallel fold to that of a similar fold (Fig. 8.2).

Few of the crystalline schists present the same simplicity of structure that one observes in slates. The movements which they have suffered may have been greater, their metamorphism has certainly been more intense, and recrystallisation has played an important role in their

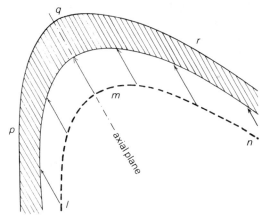

Figure 8.2 Parallel folds (folds with a constant orthogonal thickness), formed by flexural slip, may lose their parallel geometry due to flattening.

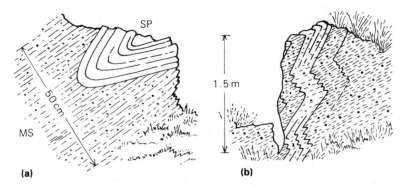

Figure 8.3 The relationship of crystalline schists to folding. (a) Axial-plane schistosity of garnet–mica schist, near Borgie, Sutherland, Scotland. MS, mica schist; SP, semipelitic. (b) Folded foliation and axial plane schistosity in garnet–hornblende schist. Ross of Mull, Scotland.

development. Nevertheless, in principle the slates and the crystalline schists have much in common. For instance, in some localities, the schistosity of crystalline schists shows the same relations to folding as does the cleavage of fine-grained slates, under these circumstances the schistosity is parallel to the fold axial planes; the relationship between stratification and schistosity can still be made use of (Fig. 8.3). Commonly however, in rocks which have suffered high-grade metamorphism, axial-plane schistosity is not present. The schistose surfaces then accord with or reflect stratification planes or bedding laminations, and continue more or less uninterruptedly around the fold hinges. In this way, the rock as a whole assumes a foliated appearance, as in a gneiss, in which the schistosity is throughout parallel to the banding.

The schistosity of a slate, to my mind, suggests the petrified laminar flow of a viscous material; the structure of a schist suggests that the flow is about to become, or has become turbulent. The main surfaces of schistosity are still prominent, but they are no longer smooth; they are, to a greater or lesser extent, irregular or corrugated.

The development of small-scale folds in schistose rocks has been discussed by Fourmarier in a number of papers (1951, 1952b, 1953a). He subdivides the structures into micro-folds (*microplissements*) and diminutive folds (*plis miniscules*). The former are corrugations of the schistosity; the latter are small-scale crumplings (*chiffonage*) of thin resistant strata enclosed in a host-rock which is schistose.

Micro-folding is considered to develop directly from normal planar schistosity as the intensity or metamorphic conditions increase with depth of burial. The transition occurs in the vicinity of '*le front inferieur de schistosité*' (the lower cleavage front) roughly corresponding to the level of Grubenmann's epizone. Fourmarier has described how, at depth, the normal schistosity gradually becomes corrugated into folds — small-scale zig-zag or sinusoidal plications — the formation of which is directly connected with the agents that were responsible for the schistosity. Rocks, in which micro-folding has developed, tend to cleave into slices whose thicknesses are controlled by the size of the plications; the orientations of the cleavages of the minerals responsible for the schistosity play a secondary role. In appearance, micro-folding closely corresponds to the British idea of strain-slip cleavage; and Fourmarier's illustrations would, I think, be referred to by that term in Britain: compare Figure 7.8 with Figures 1, 3, 4 and 5 in Fourmarier (1951).

Fourmarier's *microplissement* is confined to the zone where the coarser crystalline schists are grading towards gneisses, and it has a definite genetic significance.

Plis miniscules differ from *microplissement* in being folds of relatively thin strata enclosed in rocks that have a planar schistosity. The folds are commonly tight and drawn out and their axial planes conform to the schistosity of the host-rock. They closely correspond, on a small scale, to the 'parasitic folds' described by de Sitter (1958). These *plis miniscules* can be clearly seen in rocks of low-grade metamorphism, and even in the coarser crystalline schists they can be distinguished from *microplis*. Under conditions of high-grade metamorphism, however, the two types of folds may easily be confused.

The planes of schistosity of nearly all the crystalline schists are plicated: that is to say they are traversed by one or more sets of small-scale undulations or corrugations, the hinges of which form lineations on the schistosity. Linear features on the schistosity or flow-cleavage plane may arise by different mechanisms. The linear structures may be parallel to the direction of maximum extension as in a slate, and formed by the elongation of minerals or the stretching of inclusions. In most cases, one finds that the symmetry axes of the minor structures accord with that of the major structure on which they are dependent (Kranck 1960, Collomb 1960).

The common forms of lineations that are parallel to the hinge of major folds are:

(a) The lineation caused by the intersection of bedding and axial-plane cleavage.
(b) The schistosity may become corrugated by folding. The hinges of the microfolds may be parallel to the hinge of the major fold.
(c) Crystallisation of acicular minerals, such as amphiboles, in parallel orientation.

(d) Stretching, possibly combined with rolling, of minerals such as garnets (Fig. 8.4), or of inclusions, as for example, pebbles.
(e) The slicing of thin brittle beds into elongated strips which are parallel to the intersection of the bedding and the schistosity (Fourmarier 1949a, Fig. 328, p. 661). The final product is sometimes referred to as an 'edgewise conglomerate' (Fig. 12.3b).
(f) The development of coarse linear structures, such as drag-folds and parasitic folds, boudinage, mullions and rods. These structures are discussed in the sections which follow.

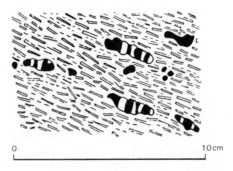

Figure 8.4 Lineation in b. Garnet–hornblende schist showing parallel arrangements of hornblendes, and stretched garnets (black) cemented by quartz (white). Ross of Mull, Scotland.

Once the rocks have developed a good axial-plane cleavage or schistosity the folds may continue to amplify by homogeneous flattening. Another method by which the folds might continue to amplify once an axial-plane fabric exists is by slip along the planes of schistosity. Folds formed this way would be termed **shear-folds**. If the movement parallel to the schistosity was laminar slip that varied systematically across a fold profile, the resulting fold geometry would approach that of a similar fold with no associated parasitic folds. This, however, is not always the case and intense movement may result in a rucking of the planes of schistosity into small monoclinal folds. This post-schistose movement may be controlled by the slip of the envelope rocks of a fold upwards towards anticlinal hinges. The frictional drag of the strata is thus transmitted to the planes of schistosity, particularly where the cleavage is more or less parallel to the fold limbs. Under these circumstances the minor folds of the schistosity will show the same sense of slip as that shown by drag-folds (Fig. 8.5a). But, where the fold has grown by a forward movement of its core relative to the envelope rocks, any rucking of the schistosity will be in the opposite sense to that observed in common drag-folds (Fig. 8.5b). In either case, the

Flow cleavage, schistosity and lineation 69

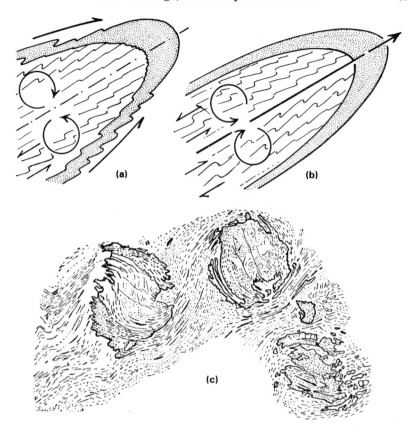

Figure 8.5 Rucking of schistosity as a result of (a) frictional drag by the envelope rocks of a fold and (b) forward movement of the fold core. (c) Rotated garnets (after Krige 1916, Read 1949b).

differential slip on the planes of schistosity will be opposed on either side of the fold axial planes. In rocks which have been subjected to large, horizontal movements, these differential effects of folding may have been suppressed by the regional translation, and the sense of the movement will be the same throughout the whole pile.

The prolonged differential slip may also produce a torsional movement between the planes of schistosity, as shown by the circular arrows in Figure 8.5a and b, and pre-existing minerals may be rotated by it. If the movement takes place while porphyroblasts are growing, spiral or snow-ball crystals, commonly garnets, may result (see also Flett 1912, Pl. IX, Fig. 6 & p. 111). The axis of rotations has been found to lie parallel to the local B-direction (McLachlan 1953) and specimens have been described in which the sense of rotation, as indicated by the garnets, was opposed on opposite flanks of small folds. Attempts to unroll spiral porphyroblasts have been made by Schmidt (1918) and Becke (1924), and it was estimated that, for a thickness of 1 cm across the schistosity, the differential movement varied from 3 cm up to 5.6 cm. The significance of these figures has been discussed by Read (1949b, p. 116). It should be mentioned here, that Collette (1959) did not agree with this mechanical origin of 'helicitic' porphyroblasts,* and considered that they could be explained by the theory of elastic anisotropy; and hence 'that these structures cannot be considered witnesses of shearing movements parallel to the schistosity plane' (Collette 1959).

Throughout this chapter, the structures discussed have been those found in regions of single phase deformation, in which the structures as a whole have been homoaxial; the b-direction has been constant throughout. Such simplicity, which results in cylindrical folding, is not uncommon, and many structures closely approximate to it over considerable areas.

In some regions, however, though the folding may at first sight appear relatively simple, it has been found that the linear structures associated with it are not parallel to each other, nor to the fold-hinges. This may be the result of superimposed tectonics; it can, however, also occur in areas where there has been one phase of movement, but where the folding was conical (Stockwell 1950). A good example has been illustrated in Clifford *et al.* (1957, Fig. 3). Here the b-linear structures converge down their plunges towards the apex of the cone formed by the folded beds.

In North Devon, my colleague Professor D. J. Shearman has shown me other anomalies between the large-scale and small-scale structures. These occur in an area where the strata are folded into asymmetrical elongated domes. This has led to the generation of oblique movements and stresses in the folded beds, so that the linear structures are not only deflected, but also do not necessarily accord with the major fold axes.

In general, it is reasonably safe to assume that the flow cleavage or schistosity of slates is an axial-plane cleavage, accordant with the major structure, as outlined in Chapter 6. The schistosity of many of the crystalline schists behaves similarly, and commonly the corrugations found on the surfaces of schistosity form lineations sub-parallel to B (the major fold axes), though there are marked exceptions. The deeper seated

---

* Read (1949b, p. 112, 1957) considered that 'if we propose to use the term **helicitic**, we had best do so in the same sense as *Weinshenk*, as indicating a texture presented by inclusions, the trails of which are relics of a folding earlier than the crystallisation of the porphyroblasts — helicitic has a time-significance that . . . should be preserved.' Syntectonic spiral or twisted porphyroblasts, such as those illustrated by Krige (Fig. 8.5c) and Flett (1912) should not, according to Read, be termed helicitic.

metamorphic rocks may have suffered more complex movements, or their structures may have been distorted by superimposed tectonics (see Cosgrove 1980). Consequently, their minor structures must not only be studied with care, but conclusions must be drawn from them with circumspection. It is only after their relationships to the tectonics have been established in one or more sub-areas of a region, that they can be used in the unravelling of the major structures.

9 Boudinage

Much has been written on boudinage structure since it was originally exhibited to *la Société Géologique de Belgique* in 1908, and later in 1922 to the *Congrés Géologique Internationale*, at Bastogne by Professors Max Lohest, Stainier and Fourmarier; but until about 1950 it was still considered a rarity. Now it is well known, but one finds that the term is being misused. It is not uncommon for any detached, freely-floating rock-inclusion in a metamorphic complex to be referred to as a **boudin**, regardless of its shape or origin. Rast (1956) has proposed the term **tectonic inclusion** for such bodies; MacIntyre (1951) called them 'fish'.

The type boudins of Bastogne were described in Lohest, Stainier and Fourmarier (1909), and their original description has been quoted or translated by numerous authors including Holmquist (1931) who added to it; Wegmann (1932) who also gave Holmquist's additional observations; Read (1934); and E. Cloos (1947a) who summarised the examples of boudinage known at that date. Quirke (1923) also described briefly the structures seen at Bastogne; but his theories of their origin by compression parallel to the layer are not generally accepted nowadays.

The most important fact, which emerges from these descriptions, is that the boudins, when they can be seen in three dimensions, form '. . . a series of large cylinders or sausages lying side by side' like sausages in a frying pan or like the fingers of one's hand when it is lying flat on a table. Unfortunately, it is uncommon that the true lengths of boudins can be observed, as at Bastogne. Nearly always one sees them in cross section: Corin's *chaplet d'osselets* (Corin 1932; illustrated also in Read (1934, p. 136)). These are usually exposed on transverse joint-surfaces, and the true cylindrical lengths are hidden, or eroded.*

The boudins of Bastogne are approximately equidimensional in cross section: '. . . having the form of a barrel the two ends of which are marked by quartz veins . . .' The majority of those described from elsewhere show marked differences in their relative transverse

---

* Confusion with regards to the significance of the term **boudin** has undoubtedly arisen in some countries because of the different ways in which sausages are displayed in shops. On the continent of Europe, large boudins are found lying side-by-side on grocers' slabs; in Britain and America, the smaller type of sausage is more common, and these are seen hanging in strings, end to end. Transverse sections of non-equidimensional boudins remind the unwary of the latter. This misinterpretation of Lohest's original description has unfortunately been made in at least two papers.

Boudinage

dimensions, and it seems to me that some standard terms for measurements would be useful. The following suggestions are largely based on Wegmann's (1932) terminology (Fig. 9.1).

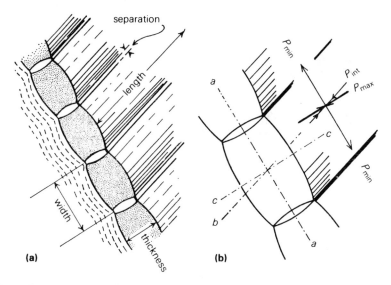

Figure 9.1 (a) Boudin dimensions and shapes. (b) Structural symmetry and stress distribution around a boudin.

(a) **Length:** measured parallel to the axis of the cylinder, or to the line of junction between two adjacent cylinders. The term elongation in this sense is misleading and *should not be used*, because the extension or elongation of the rock-mass as a whole is normal to the length of the boudin.
(b) **Thickness:** measured in the same manner as one would measure the thickness of a bed or dyke.
(c) **Width:** the distance, measured at right angles to the axis of the cylinder and parallel to the original plane of stratification, across an individual boudin, that is, from edge to edge across the cylinder.
(d) **Separation:** the distance, at right angles to the cylinder axes, between adjacent cylinders. **Longitudinal separation** may be used to measure the gaps formed across the main cylinders where they have been broken by a transverse boudinage system, such as would yield a 'tablet of chocolate' structure (Wegmann 1932).

The typical boudin has been derived from a bed or dyke which was more rigid than the material which enclosed it, at the time when the structure

Small-scale geological structures

Figure 9.2 (a) Cross sections of slightly rotated boudins, Moine Series, Ross of Mull, Scotland.

was formed. Its transverse section may vary from the nearly equidimensional barrel-shaped boudins of Bastogne, to that of a flat lens. Commonly the sides of the boudin are parallel in their central portion, but they converge towards the edges, which themselves may be concave. The less rigid beds (or rocks) on either side conform to the shape of the boudin, and curve into the constriction (or neck) between adjacent boudins (Fig. 9.2b). Depending on the circumstances the boudins may be separated by lenticular segregations of quartz or calcite, or by country rock which has been squeezed into the gap between them.

Boudinage 75

Figure 9.2 (b) Attenuation and boudinage on the flanks of an upright fold composed of impure quartzite and mica schist layers, Moine Series, Ross of Mull, Scotland.

The structure as a whole suggests a drawing apart of the rocks in a direction at right angles to the lengths of the boudins. In appearance, the constriction (or neck) between adjacent boudins is analogous to that which develops in a mild-steel test piece under tension, as it approaches the point of fracture.

This analogy must not be carried too far, because the conditions are not quite the same. The test piece is broken, in the free air, by tension only; boudinage occurs at depth in the Earth's crust, where the surrounding pressure resulting from the weight of superincumbent rock, and from tectonic forces must be taken into account. Ramberg's (1955) experimental work has demonstrated that boudinage does not necessarily represent the action of a direct tensional force, but of a stress difference, $P_{max}-P_{min}$. His model consisted of a layer of brittle material lying between two layers of more plastic material – like a sandwich. On compression normal to the layering, the exterior plastic layers flowed outwards, while the more rigid bed between them failed by tension, because of the frictional drag on its upper and lower surfaces. This has been supported by further experimental work in which natural rock material was used (Griggs & Handlin 1960). Rods of relatively brittle rock were encased in cylinders of more plastic rock, which in turn were subjected to high confining pressures at various temperatures. The apparatus was arranged so that the central rods could extend longitudinally as the surrounding pressures were increased. This longitudinal extrusion of the test pieces was controlled, and various types of rupture were observed. The forms of these breaks depended on the ductility of the core. Brittle material showed a sharp fracture at right angles to the axis of the test piece; less rigid rods showed typical constrictions before they parted, and the surrounding plastic envelope was forced into the constrictions in a manner similar to that seen in natural boudinage.

Paterson and Weiss also produced boudins in natural rock. The boudins developed in quartz veins in a slate. The orientation of the boudins thus indicates the local directions of the principal stresses (Fig. 9.1b), P_{min} is at right angles to the axis of the boudin, and parallel to its width; it corresponds to the direction of extension of the rock mass as a whole. The length of the boudin is parallel to P_{int}. The structure itself is

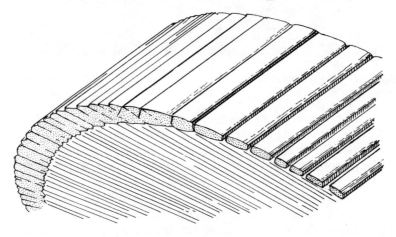

Figure 9.3 Boudinage on the flanks of a fold.

orthorhombic and indicates the directions of compression, of extension, and − parallel to the lengths of the cylinders − of no change.

The normal orientation of the lengths of boudins is thus parallel to the principal tectonic axis, or axis of folding. This was observed at Bastogne (Quirke 1923). E. Cloos (1946, 1947) McIntyre (1950) and Sanderson (1974) noted the same thing; and the lengths of the boudins in the two examples shown in Figures 9.2a and b are also parallel to the local and major fold-axes; the b direction of the boudins = B. Wegmann (1932, p. 484) has pointed out that 'Recumbent folds seem particularly favourable for the development of boudinage' and Gindy (1953) illustrates an example of this.

Boudinage is commonly associated with folding, in which the development of schistosity in the more ductile beds has been accompanied by a plastic attenuation of the more rigid strata on the fold-flanks (Fig. 9.3) and McIntyre (1950). This stretching may even continue into the hinge-zone of the fold (Fig. 9.2b) (Gindy 1953, Rast 1956), though not necessarily always.

The attenuation of the reversed limb of a recumbent fold may also, as Wegmann noted, be accompanied by the formation of boudinage. Similarly it may be present in the vicinity of important thrusts (Fig. 13.2c) as at Tintagel, in Cornwall (Wilson 1951, Pl. XXIX, 2). Here, in the cliffs of Trebarwith Strand, are boudins which measure between 1 to 1·5 m in thickness, and are up to 6 or 7 m in width. One was observed to have a length of over 20 m. Parkinson (1903) described smaller-scale structures from the same locality as '. . . a series of three or four ellipses joined the one to the other like the links of a chain'. Throughout this area the lengths of the boudins are parallel to the tectonic b direction.

The lengths of boudins are not necessarily parallel to fold axes. The production of boudins, whose lengths are oblique to the local fold axes, has been discussed by Rast (1956). He mentioned that examples occur in the Dalradian of Scotland; but he considered that the evidence as to their origin was inconclusive. Professor D. J. Shearman, working in North Devon where the strata are folded into asymmetrical elongated domes, has found that the lengths of all the boudins are oblique to the axes of the folds. They are, however, parallel to the local tension gashes. This suggests that the local stresses in the rocks themselves were oblique to the external tectonic forces responsible for the folding. Read (1934, p. 135) described and illustrated an example of a boudinaged dyke in a zone of thrusting, in which the lengths of the cylinders were parallel to the direction of movement. The dyke was enclosed in soft graphite-schist which had apparently been formed during the movement of thrusting (Fig. 9.4). Cloos and Hietanen (1941, p. 155–7) noted that they had

Small-scale geological structures

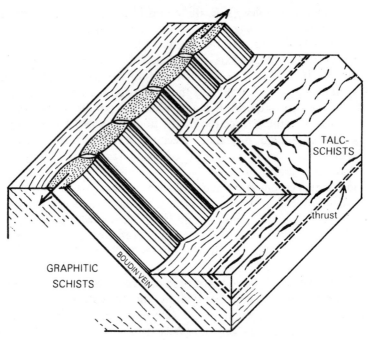

Figure 9.4 Boudins in *a* parallel to the direction of movement on a thrust zone (after Read 1934).

found boudins oriented both parallel to, and normal to the local tectonic axis in the same quarry. Wegmann's *tablette de chocolat* structure, also observed at Bastogne, has resulted from the intersections of boudinage developed in two directions at right angles (Wegmann 1932, Coe 1959).

Rotation of boudins about their longitudinal axes can in some places be recognised. This converts a simple orthorhombic structure into a monoclinic one. Normally, a flat-sided boudin is prevented from rotating by the beds on either side; but, if the boudin be equidimensional in cross section, or if the rocks are undergoing a strong shearing movement, it may be partially rolled. In folded strata, the sense of the rotation will conform with the slip between the beds – upwards towards anticlinal hinges. Hence the asymmetry of these minor structures can be used as an aid in determining the form of the major structure. For example, in Figure 9.2b the slight twist of the boudins in a clockwise direction is in accord with the slip of the beds on the major structure. The exposure illustrated lies on the steep (vertical) limb of a big asymmetrical syncline, the hinge of which lies to the left of the photograph; and the beds to the left have moved upwards, out of the syncline, relative to those on the right (Fig. 13.2a).

E. Cloos (1947a) has illustrated an example in which almost equidimensional boudins have been rotated through some 40°. He also discusses the mechanism of rotation. These boudins of Cloos' have rhombic cross sections, not the characteristic boudin form. It is possible that the rigid bed did not originally fail by tension, but by shear; the prismatic cylinders might, perhaps, be equidimensional fracture-cleavage slices, rather than blocks separated by tension; the effect of the rotation is nevertheless the same. Boudins with rhombic cross sections and the rotation of boudins are discussed by Stromgard (1973).

The development of boudinage in gneissose terrains has been described by Wegmann (1932) and Ramberg (1955). Both these workers have demonstrated magnificent examples of basic layers or dykes in banded gneisses being pulled apart by the movement of the more acid material which surrounds them. The contrary was observed by Scheumann (1956), who has made a detailed study of small-scale complex boudins in a strongly deformed plagioclase-amphibolite, in which the movement was one of viscous flow. In general the leucocratic plagioclase-rich layers were segmented to form the boudins, while the surrounding melanocratic material flowed in between them.

The stretched belemnites described by Heim (1878, Atlas, Pl. XV, 1921, T. II (1) Pl. III) must be considered a linear variety of boudinage. The fossils have been stretched and broken into segments which are now cemented together by quartz. Because of their original cylindrical shapes, their 'lengths' and 'thicknesses' are the same. Their extension is known to be in a. A more recent study of stretched belemnites has been published by Hossain (1979). Similar stretching has been observed in garnets and in pebbles which have been segmented and the space filled in by quartz. The elongation in some cases was in a (Clough 1897) and in others in b (Fig. 8.4).

In brief, boudinage structure is normally an orthorhombic structure in which the lengths of the cylindrical boudins declare the direction of the local intermediate stress, or direction of minimum distortion. Their widths indicate a direction of stretching or the local minimum principal stress. Boudins are thus commonly oriented parallel to the tectonic direction, in b, but also occur normal to it, in a. Oblique boudins are rare, but have been observed. Rotated boudins form monoclinic structures, and the sense of rotation indicates the relative sense of the slip between the rocks on either side of the boudinaged bed or dyke.

10 Drag-folds and parasitic folds

Drag-folds, literally *plis d'entraînement*, but termed by Moret (1947) *plis de frottement* (friction folds), are asymmetrical folds produced in the weaker beds by differential movements of the more resistant rocks above and below them. They are disharmonic folds, usually confined to definite zones or groups of rocks. The weaker strata, being unable to resist the frictional stress of effect induced by the movements of stronger beds above and below them are dragged into these overturned or asymmetrical minor folds, the axes of which lie at right angles to the direction of movement. The sense of the movement is indicated by the inclination of the axial planes or the *regard* of the folds: they are typical monoclinic structures (Fig. 3.1c).

In size, drag-folds may vary between wide limits, from small plications in thin beds, to folds one of which may form the whole face of a mountain – as for example, some of the autochthonous folds of the Alps.

The most common setting for drag-folds is on the flanks of major folds; they may also occur in regions of thrust tectonics and where the gliding of an approximately horizontally stratified rock-mass has taken place. Rarely they may be found in the vicinity of normal faults. They denote a slipping movement parallel or subparallel to the stratification.

In an area of simple flexural folding, the drag-folds are predominantly the result of the slip on the planes of stratification produced while the major structures were being formed. As has already been pointed out, such slip results from the upper strata moving upwards towards anticlinal hinges, relative to the lower strata (see Ch. 4). The *regard* or asymmetry of the minor plications declares the sense of this relative movement between the various beds; examples are given in numerous publications, (Leith 1923, Nevin 1949, Hills 1953, Billings 1942, Wilson 1946). In a simple fold having a horizontal axis, the movement will be directly up the dip of the beds: the plane of symmetry will be vertical, and the axes of the minor folds will be horizontal, parallel to the axis of the major fold. If the major fold is plunging, the axes of drag-folds will plunge likewise. Consequently by plotting the orientations of the axes of the minor folds, one may obtain a picture of the behaviour of the major structure on whose flanks they lie (Fig. 10.1).

Drag-folds and parasitic folds

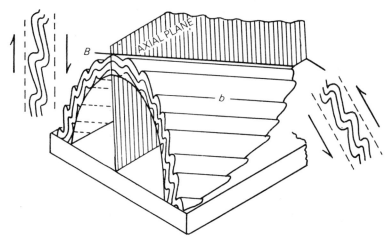

Figure 10.1 Drag-folds on a major fold, showing parallelism of plunge.

In regions where the major folds are overturned both limbs dip in the same general direction; and the drag-folds, like cleavage, can be used to distinguish normal (right way up) limbs from those which are inverted. If the upper beds have moved upwards, one is on the normal limb. If the asymmetry of the minor folds indicates that the *lower beds* have relatively moved upwards, the strata are inverted, and one is on the inverted limb of the fold (Fig. 10.2a).

When one is working on poorly-exposed ground, a reversal in the sense of movement of gently plunging drag-folds commonly indicates that one has crossed the hidden hinge of a major fold.

In regions where shear folding, rather than flexural folding, is dominant, the development of minor folds on the flanks of major folds becomes more complex. Such minor folds are no longer true drag-folds, though they may have been initiated by the slip on stratification planes in the early stages of the movement. With the formation of an axial-plane cleavage, and the growth of the fold by compression normal to, and extension parallel to the cleavage, the original minor flexures likewise become amplified. Their forms are too drawn out to be the result of drag alone; and de Sitter, adopting a term proposed by Professor D. J. Shearman, has referred to them as **parasitic folds** (de Sitter 1958, p. 280). He, de Sitter, has also demonstrated that these parasitic folds will only develop in true shear folds when the limbs are at an angle of over 45° to the axial plane. He thus has accounted for the presence of strong minor folding or crenulations in the hinge region of folds of this type. If the limbs of the fold are closed to less than 45° to the axial plane, they

82　Small-scale geological structures

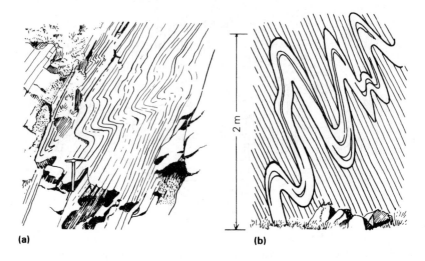

Figure 10.2 (a) Drag-folding on the overturned limb of a fold, Kopaonik Mountains, Jugoslavia. (b) Isoclinal parasitic folds of a semipelitic schist band in mica schist. A syncline is known to lie to the left of the sketch and an anticline to the right. Ross of Mull, Scotland.

will no longer fold, but become attenuated or even boudins, as shown in Figure 9.3. Because many large folds are formed by a combination of flexure-folding and shear-folding, it is probable that many parasitic folds may likewise be produced by both mechanisms working together or in sequence. This results in the production of minor folds which are compressed and drawn out, but which can still be used by the field geologist in the same way as simple drag-folds (Figs 10.2b, 13.1).

Minor folds which accord with the major-fold structure are said to obey 'Pumpelly's Rule' (Pumpelly, Wolfe & Dale 1894),* and have been termed **dependent drag-folds** by Derry (1939). **Independent drag-folds** are those which are the product of some secondary movement, and so are not parallel to the main tectonic axes. Hills (1940) used the terms **congruous** and **incongruous** in the same sense.

One common cause for the production of independent drag-folds is a horizontal movement parallel to the planes of stratification in beds which are already steeply inclined. Such movement may occur after the main regional folding has taken place, and the more or less vertical strata become dragged into folds which have steep axial plunges (*Steilaxen*).

---

* The original statement in Pumpelly *et al.* (1894) reads: 'The degree and direction of the pitch of a fold are often indicated by those of the axes of the minor plications on its side.'

The same mechanism may also result in the formation of a steeply inclined cleavage in the rocks. Derry (1939) discussed the importance of these independent drag-folds in some of the mining districts of Canada. Large-scale examples have been recognised in the Austrian Alps (*Schlingentektonik*) by Schmidegg (1933), Metz (1957); in Canada by Wynne-Edwards (1957); Ruhland (1958) in the Vosges; and by Lillie (1961) in New Zealand. Under these circumstances the plunges of the structures are controlled, not only by the direction of the movements, but also by the dip of the beds affected.

Drag-folding of strata is also commonly associated with large thrusts. The minor fold-axes are parallel to the *b*-direction of the major structure, and the asymmetry or *régards* of the folds declare the sense of movement on the thrust (Wilson 1951, Pl. XXIX, 1). They may also be developed below such structures as nappes, where the movement of the upper tectonic units has acted as a *traineau écraseur* (crushing sledge) on the rocks below. The autochthonous folding in the Alps can be cited as an example. Overturned drag-folds occur in the Cambrian strata below the Moine Thrust in north-west Scotland; there, because of the relatively brittle nature of the rocks, they rapidly grade into minor thrusts and imbricate structure. The direction and the sense of the Postcambrian movement on this great dislocation is shown not only by these structures, but also by the stretching of pebbles and by the bending of annelid tubes − like a candle flame in a draught (Peach & Horne 1907, pp. 495 & 499).

Similar structures are also found in regions of gliding tectonics, where the main movements have been the even sliding of the rocks along planes of stratification or of schistosity. Locally, this even movement has been interrupted by the surfaces becoming corrugated and asymmetrically folded (Fig. 13.2b and c). Under ordinary conditions these minor folds and corrugations are oriented at right angles to the main movement; but in rocks where the deformation was one of shear-folding it is quite possible that the symmetry of the minor structures is not in accord with that of the major tectonic movement (Ramsay 1960).

Naturally, in regions of thrust tectonics and of gliding tectonics, although the orientations of the minor folds can be used to determine the direction and sense of the main tectonic transport, they cannot be used to elucidate the stratigraphical succession of the beds as in simple and overturned folds.

Individual drag-folds and parasitic folds, when traced along their lengths − in North Devon they are popularly known as 'fossil trees' − behave like ordinary folds. They are more or less cylindrical in their central portions (Fig. 5.2) but die out or become conical at their ends (Fig. 10.3). In consequence the plunges of individual minor folds, when

Small-scale geological structures

Figure 10.3 Parasitic drag-folds crossed by tension gashes and plunging parallel to the major structure. The dying out of each fold along its length indicates that they are in reality pod-folds.

seen in profile, may not always be parallel to that of the major structure; but if the plunges of several minor folds be measured, the statistical average will, in general, closely accord with the plunge of the major structure.

One type of parasitic fold which has been recognised, and which may be misleading, has been encountered in the Roan Antelope copper mine (Mendelsohn 1959). This structure has been termed a **pod-fold**, from its shape which resembles that of the seed-pod of a pea or bean (Fig. 10.4). Similar structures have been recognised by Campbell (1958), who called them **pointed elliptical folds**.

Structures identical with Mendelsohn's pod-folds, but on a larger scale, have been recognised in the Moine Series by Sutton and Watson (1954); they were not given any special name, but were considered to be a variety of **inconstant fold**. The anatomy of these structures, and their tectonic significance are described on p. 38 *et seq.* of the paper referred to above; and the clearest example, which has a length of approximately 4 km is that at Aultdearg (*loc. cit.* Pl. II and III).

In a pod-fold, the fold-axes are not parallel, but diverge and then converge from one end of the structure to the other. In consequence, 'the steep or inverted middle limb common to a pair of these folds (the anticline and syncline, which together form a single pod-fold) is

Drag-folds and parasitic folds

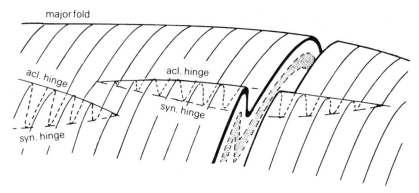

Figure 10.4 Diagram of a 'pod-fold' (after Mendelsohn 1959, Fig. 6), showing the convergence of the anticlinal and synclinal hinges.

thickened, thinned, or entirely eliminated by variations in the plunge of the two structures' (Sutton & Watson 1954, p. 38). Hence, neither the anticlinal, nor the synclinal hinges are themselves parallel to the major tectonic structure; but, as these two authors have pointed out, the mean of the two taken together accords with the regional trend.

The details of pod-folds can be most clearly seen in mining districts, where their shapes in three dimensions can be recognised in the underground workings. Mendelsohn referred to folds of this type as probably occurring at Broken Hill, Australia, and at Homestake, South Dakota. He also considered that they may be much more common than has hitherto been realised; and that they and folds *en échelon* may, in fact, be the basic fold forms (Mendelsohn 1959, p. 235).

In general, we can say that the minor disharmonic asymmetrical folds – drag-folds and parasitic folds – which are found in rocks, can normally be used as indications of the direction and sense of local differential rock-movement. In a folded terrain they commonly indicate the order of the stratigraphical succession; but in zones of thrusting this is not true. Gently-plunging minor folds are commonly parallel to the major tectonic axes. Steeply-plunging folds suggest secondary horizontal slip on beds which were already steeply inclined, or possibly the upturning of early gently-plunging folds by a later movement. Nevertheless, there are exceptions to the general rules, and care must be taken here, as elsewhere, in extrapolating from the minor structures to the major – especially when working in regions of highly metamorphosed rocks, or where tectonic movements of different ages may have been active.

11 Mullion and rodding structures

Mullion and rodding structures are both forms of a coarse lineation developed in rocks which have been strongly deformed. In general they are parallel to fold hinges. In the past the two terms were considered more or less synonymous; and Holmes (1928, p. 16) described mullion structure as recalling

> the appearance of the clustered columns which support the arches, or divide the lights of mullioned windows, in Gothic churches. The structure is also described as rodding structure, and is typically developed in the Eirebol district, where 'rods' of white quartz, varying in dimensions from those of telegraph poles to those of walking sticks, lie parallel to each other down the dip slope of the Moine Schists. Where minerals of elongated habit like hornblende or biotite are present in the rocks showing mullion or rodding structure the crystals are arranged parallel to each other and to the dip and plunge of the folds.

I have distinguished between these two types of structure; and have confined the term **mullion** to the structures which have been formed from the country-rocks themselves (Fig. 11.1); and **rods** to those which have been developed from quartz or other minerals that have segregated in, or have been introduced into the rocks while movement was in progress.

The first use of the term mullion structure is not known, but it was referred to in print by Hull, Kinahan and Nolan in 1891, who observed it in Donegal, Ireland. They employed the term as if it were one which was already familiar to them, but no earlier reference has been found. Fermor (1909) observed similar structures which he described as 'parallel striated and grooved prisms suggesting logs of wood', in the manganese deposits of India, and named them 'slickensides-grooving'. In 1924 he gave examples illustrating the parallelism of such coarse linear structures to the plunges of the local folds. 'Corduroy structure' was a term suggested for all varieties of lineation which formed ridges or undulations on the rock surface, by Bailey and MacCallien (1937, p.

Figure 11.1 (a) Steeply plunging mullions, Dalradian Series, near Portsoy, Banffshire, Scotland.

103). They included the rippling of schists, gaufrage, and microfolding etc., together with mullion structure, under this term. Leith (1923, p. 100) considered mullions as being a coarse form of slickenside striation, like the Rodadero (Gregory 1914), an idea which was commonly held by early workers in the North-West Highlands of Scotland. This theory was modified by Read in 1926, when he suggested that structures had resulted from the interaction of two deformations — the first a compression normal to the lengths of the mullions; and the second, a stretching parallel to their lengths (Read & Phemister 1926). Coles Phillips (1937, p. 597) supported Read after having studied the

Small-scale geological structures

Figure 11.1 (b) Mullions on a vertical plane of stratification, plunging parallel to the local fold axes. The mullions are covered with a thin veneer of polished manganese ore, Kandri Mine, Central Provinces, India.

microfabric of the mullions at Oykell Bridge in Northern Scotland. I visited this area in 1949, and later described the geological setting of these mullions (Wilson 1953). I considered that the second deformation suggested by Read, parallel to the lengths of the mullions, played only a very subsidiary role. Some years ago, Professor A. Pilger and Dr W. Schmidt (1956, 1957) recognised the development of mullions in the N. Eifel district of Germany, and have since made a series of most detailed studies of these occurrences. The structure has also been observed in the Caledonides of Scandinavia. Lindstrom has described examples, and has also commented on Pilger and Schmidt's work in the Eifel (Lindstrom 1958).

Mullions seem to be typically developed in strongly deformed metamorphic rocks, though those of the Eifel lie in a region of relatively low-grade metamorphism. In Scotland they are found in the Lewisian Gneisses (Peach & Horne 1907, Pls. XXV, XXVI), and in rocks of biotite or garnet grade. Coles Phillips (1937) demonstrated that the lengths of the Oykell Bridge mullions were parallel to b; and I found them to be parallel to the local fold axes, $b = B$. Since then I have mapped mullions elsewhere in the Moine Series of Scotland and in India, and have found no reason to change this opinion. Nevertheless, Pitcher and Read (1959) have illustrated beautiful mullion structures in the wall-rocks of the Main

Donegal Granite, and considered that they are aligned parallel to the direction of movement of the intrusive rock, that is, in a. Lindstrom (1958) has illustrated structures to which the term mullions might well be applied. These examples lie in a region of thrust tectonics; and are considered to be oriented more or less parallel to the principal direction of movement. Kvale also suggested that the mullions of Oykell Bridge, and much of the lineation seen in the Moine Series of Scotland, were likewise aligned in a (Kvale 1953).

In shape, mullions are long cylindrical structures of which the surfaces may be rounded, or they may be irregular, like a sheet of corrugated iron which has been rolled up longitudinally (Fig. 11.1). The external surfaces may be polished, covered by a thin veneer of mica or some other mineral, or they may be striated parallel to their lengths. Internally, the rock of which mullions are composed is solid and massive throughout, even though it may be laminated. Commonly they are cut by cross-joints at right angles to their lengths; hence they break away from the exposure like sections of fluted pillars (Wilson 1953, Pl. 7, Sander 1948, Figs 44 & 45). In size they vary from cylinders or irregular prisms some 2 cm or 3 cm in radius or less, to curving surfaces which have radii of 2 m or more, so that they look like partially buried water mains. Locally, as in Donegal, mullions have been used as fence posts or gate posts; and at Oykell Bridge one stands as a monolith some 3 m in height above the ground.

Mullions can be classified into three varieties:

(a) fold-mullions or bedding mullions;
(b) cleavage-mullions;
(c) irregular mullions.

Many are formed by a combination of two or three types.

Fold-mullions possess regular curved cylindrical surfaces which correspond to original bedding or to pre-existing planes of foliation. Commonly they are largely composed of detached or strangled hinges of parasitic folds, and the bedding lamination within the mullion accords with the external surface, (Wilson 1953, Fig. 3, p. 125). The 'waterpipes' of the River Garry in Scotland are likewise formed from the hinges of sharp folds (Barrow *et al.* 1905), which are not necessarily detached.

Bedding-mullions are undulations of the bedding plane surfaces which have been smoothed, polished, or striated. Locally they may be formed by a single bed pinching and swelling, elsewhere they may be gentle flexures or large corrugations (Fig. 11.1b). In the manganese mines of the Central Provinces, India, such mullions are coated with a highly polished veneer of manganese oxide, and one fold mullion had a continuous length of 100 m; all were parallel to the plunges of the local fold axes.

Cleavage-mullions are long rock-prisms which may be more or less angular, or partially rounded in cross section (Wilson 1953, p. 126, Pilger & Schmidt 1956, 1957). The prism surfaces are dominantly cleavage planes, and commonly these have been folded by further movement. One or two of the sharp edges of the cleavage slices may be ground away, so that one is left with a cylinder having an approximately oval cross section. Other varieties may show a curved face on one side, while the other surfaces are relatively flat. All, however, are characteristically polished, mica-covered or striated.

Irregular mullions are the most common variety. They are long cylindrical structures, but in cross section they are very irregular, and interlock, each one with its neighbours, like pieces in a jig-saw puzzle. The cylindrical surfaces are grooved, like worn cog-wheels, and striated or covered with a micaceous veneer (Fig. 11.1a); (Wilson 1953, Fig. C, p. 126). The internal structure of the mullion may show contorted bedding laminae, and these may locally accord with the external surface but for the most part they are truncated by it, as noted by Coles Phillips (1937, Fig. 6, p. 596).

All these various types of mullions may occur together, in any one area, and they are strictly parallel. When I have seen them they are cylindrical *B*-structures (Sander 1948, Figs 44 & 45), and result from movements normal to their lengths; but they will not declare the sense of the movements, as they are not clearly monoclinic, when seen in the field.

The manner in which these structures are so clearly developed, their very strong lineation, and in some cases the marked mineral orientation which occurs in them (the *c*-axes of hornblendes, for examples, are parallel to the axes of the cylinders) all suggest that, in addition to a rotational or a compressional deformation, the mullioned rocks have also suffered a concomitant stretching parallel to their lengths. I thus disagree with Read's suggestion of two separated deformations; and consider that mullions are examples of *Einengung* (Sander 1948, Weiss 1954), in which the rocks have suffered a great squeezing normal to their lengths, but under conditions such that stretching parallel to their lengths was possible.

In rocks which have been cylindrically folded, therefore, the axes of the mullions are parallel to the axes of the folds. Their orientations correspond to those of the hinges of drag-folds, discussed in Chapter 10. Mullions may also form in areas where the folding is more complex as for example in folds which Professor John Sutton has referred to as

inconstant folds (Clifford et al. 1957, p. 6). There, their orientations are controlled by the local stress distribution. For instance, in the Fannich Mountains of Scotland, where the folding is inconstant, mullions and other linear structures show a steady change in orientation between the converging or diverging regional fold-axes (Sutton & Watson 1954).

Steeply-inclined mullions, like drag-folds which show steeply-plunging axes, may also be formed in regions where more or less horizontal movements have taken place along steeply dipping bedding or foliation, and are characteristic *steilaxen* (Fig. 11.1a).

Rods are cylindrical bodies of quartz or other minerals which have segregated in, or have been introduced into the country-rock while the tectonic movements were going on. They differ from mullions in that they are not composed of the country-rock itself, but are essentially monomineralic. The most common mineral to form rods is quartz; rarely they may be developed in calcareous rocks and are then composed of calcite; in the Department of Geology, Imperial College, we have a rod of pyrite, from the Rio Tinto mining district of Spain.

The classic locality showing rods in Britain is on the mountain of Ben Hutig ('Beinn Thutaig' on Sheet 114, Geological Survey of Scotland, and in the early reports) in north Sutherland, Scotland (Peach & Horne 1907, p. 603; Wilson 1953, pp. 131–8). Here one can see rods of vein quartz which vary from about 60 cm to about 1 cm in diameter. They '. . . are composed of irregular grains of quartz with some flakes of white mica. The rods are associated with mica-schists of Moine type, which have been denuded into hollows, these siliceous ribs project and form conspicuous features' (Peach & Horne 1907, p. 603). Some of the dip-surfaces of exposures look as if they were covered with a layer of parallel white bamboo canes; a few of these may be as thick as a man's arm. Nearly all the bigger rods are striated parallel to their lengths.

The more impressive of the Ben Hutig rods lie in the hinge-zones of parasitic folds. The quartz has segregated in these zones of relatively low pressure; probably as Hallimond suggested, as miniature saddle-reefs (in discussion of Wilson 1953, p. 146). Then, as the movement continued the segregations were rolled up at right angles to their lengths (Fig. 12.4a).

Other rods have originated from the deformation of quartz veinlets which were parallel to the foliation or stratification of the rocks, and which were then folded with the material surrounding them (Fig. 11.2a). The resistant quartz hinges now form long projecting ribs which one sees on the weathered rock surfaces. The deformation of early tension gashes which were oblique to the planes of movement may likewise result in the formation of isolated rods of quartz (Wilson 1953, Fig. 9).

In transverse profile, the rods of Ben Hutig vary from oval or nearly

92 Small-scale geological structures

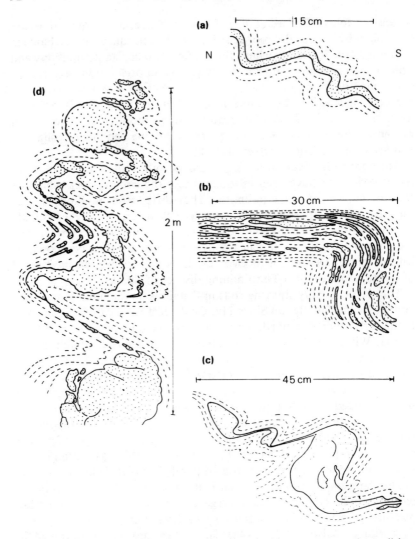

Figure 11.2 Profiles of quartz-rods from Ben Hutig, Scotland. (a) Quartz vein parallel to, and folded with the stratification. (b) Stretched and folded quartz-lenses parallel to the stratification, formed from conglomerate pebbles. (c), (d) Larger quartz rods developed from segregation quartz in the hinge zones of parasitic folds (after Wilson 1953).

circular, to very irregular cylinders or lenses. The extent of the rounding depends on the original shape of the quartz body, and on the amount of movement which the segregation quartz has suffered.

More recent work in this area has shown that many of the thinner rods on Ben Hutig and in its neighbourhood are, in truth, derived from deformed conglomerate pebbles of quartzite and vein-quartz. These have suffered from two successive, but continuous phases of deformation. In the first, the pebbles were flattened and drawn out into thin elongates lenses; the elongation is parallel to b. In the second phase, these lenses were folded with the rocks that contained them (Fig. 11.2b) or they were, themselves, rolled into siliceous rods by slip. In both cases the axes of rotation, B, were parallel or nearly parallel to the original b-direction. Some of the larger pebbles which have been deformed in this manner are twisted obliquely to their lengths, which suggests that locally there may have been slight divergencies in direction between the two phases of movement. It is obvious that the rods formed by the extension and rolling of conglomerate pebbles do not conform to the definition of rods given on p. 91, which applies to those derived from segregation quartz. Nevertheless, the final products of the two may be so difficult to separate in the field, that until the original parentage of the structure can be established, the term rod, which is descriptive and has no genetic significance, may with reason be applied to both. Later it can be qualified if necessary.

Rods are thus, like mullions, parallel to the axes of the local folds, and lie at right angles to the maximum compression. They vary from orthorhombic to monoclinic structures, in which $b = B$. A similar relationship has been observed by Coles Phillips in the crystalline schists of Start Point, South Devon, where 'quartz-rods are in places well developed and accord in direction with other lineated features in these rocks' (in discussion of Wilson 1953, p. 147).

12 Superposed minor structures

Un cas remarquable est encore à considerer, c'est celui qui se produit lorsqu'un pli est lui-même replié, soit lors d'une nouvelle période de plissement, soit directement dans la même période.*

These words, written by de Margerie and Heim in 1888 (p. 61) are equally applicable to this section. In the Alps, and other young mountain ranges, the complexities of refolding can commonly be seen in the great natural cross sections formed by the mountain cliffs. Elsewhere, in the older orogenic zones, much of the structure can only be seen in two dimensions, and the third has to be deduced from surface evidence. This may be greatly facilitated if one appreciates that each successive movement will not only modify the pre-existing structures, but at the same time is liable to produce its own minor structures. If, therefore, we observe that one set of minor structures regularly deforms another we have an indication that more than one movement has taken place. Then, if we can link these minor structures with their respective major structures, we will have taken a long step in the elucidation of the tectonic history of our area.

The ultimate effects of superposed movements may vary between wide extremes. In some regions, the later movements may have been so intense as to wipe out all evidence of earlier ones. In others, careful search may reveal structures which indicate that the rocks have passed through more than one period of deformation. Thus Wegmann (1923, 1931) recognised successive phases of folding in the Schistes de Casanna, each accompanied by the formation of a new schistosity which tended to efface the pre-existing structures. In the Lewisian of Scotland, the early Scourian facies and structures (+ 2700 m years old) were almost completely transformed during the Laxfordian movements some 1000 m years later. It is only near the contacts between these two groups that one can recognise the overprinting of the Laxfordian structures onto those of the Scourian (Sutton & Watson 1951, Sutton 1960). Broughton (1946)

---

* 'An interesting case should also be considered, it is that which is produced when a fold is refolded either during a later period of folding or during the same period.'

has described the manner in which a series of rocks having been folded with the accompanying development of an axial-plane schistosity, was then again deformed. The second movement resulted in the refolding of both the folded strata and the early schistosity, with the concomitant production of a new cleavage parallel to the axial planes of the second folds. This later deformation was associated with the development of thrusts.

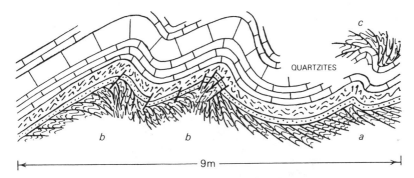

Figure 12.1 Folded cleavage in the Moine Series, Ross of Mull, Isle of Mull, Scotland. *a*, normal cleavage-bedding relationships; *b*, cleavage twisted and squeezed into anticlinal cores; *c*, cleavage curving over a tight anticlinal fold, looking southerly.

An example of superposed structures is shown in Figure 12.1 in which an early cleavage lying at about 30° to the stratification has been distorted by later folding. The cleavage indicates a movement of the upper beds to the left, relative to the beds beneath. The folding has a *regard* in the opposite sense; nor does the strike of the cleavage accord with the axial directions of the folds. The folds illustrated are drag-folds on the flank of a major syncline which lies to the left of the figure (Fig. 13.2a). The origin of the cleavage is uncertain, but the structure indicates that a pre-folding movement of the rocks had taken place.

The distortion of an early cleavage by a continuation of the same movement without interruption may also take place. In the A'Mhoine area of north Sutherland, Scotland, I have observed that the first structure to form as the result of a slipping movement of the beds, was a schistosity, more or less parallel to the stratification. This was accompanied by the flattening of conglomerate pebbles in the same plane; the pebbles were also elongated parallel to the lineation in the schistosity. Continuation of the movement resulted in the folding of the beds and the schistosity together, so that the flattened pebbles curve around the folds (Figs 11.2b, 13.2b). Locally a second, axial-plane schistosity cuts across the folded bedding and the first schistosity. The

96 Small-scale geological structures

quartz-rods of Ben Hutig in this area were formed by the rolling of the flattened pebbles and quartz segregations, during both phases of the movement (see Ch. 11, p. 91–3).

Stages in the deformation of early folds and their axial-plane cleavages by a later shear-cleavage, which in turn developed into the axial-plane schistosity of a second set of folds, have been studied by Professor J. G. Ramsay and myself in the cliffs of Holy Island, Anglesey, North Wales. Many of the interesting structures, which we observed there, are shown in Edward Greenly's classic memoir (1919) on this district. Typical examples of bedding, cleavage and schistosity (S_1) in the inverted limb of one of the first folds are seen on Figure 6.2 and in Figures 6.1a and

Figure 12.2 Early schistosity S_1 becoming distorted by later movements, South Stack Series, Anglesey, North Wales.

12.3a. An early stage in the twisting of this first structure by the second movement is illustrated in Figure 12.2. In places, the movement on the axial-plane schistosity (S_1) of the first folds had been so strong that the soft argillaceous strata were squeezed into the cleavage planes between slices of more rigid greywacke in which the original bedding laminae are still visible (Fig. 12.3b). This complex grades into a rock showing a banded foliation which is parallel to the first schistosity (S_1), but not parallel to the original bedding. The second, strain-slip cleavage (S_2) cuts this foliated rock; and as the movement became more intense, it developed into the axial-plane schistosity of the second fold-system (Fig. 12.3c). These latter folds are shear folds, *but it is not simply the stratification which has been folded, but a foliation of the first generation.* In other localities one may observe the interbedded greywacke and slates, in their normal condition, and their S_1 cleavage, being folded together by the second movement (Fig. 12.3d).

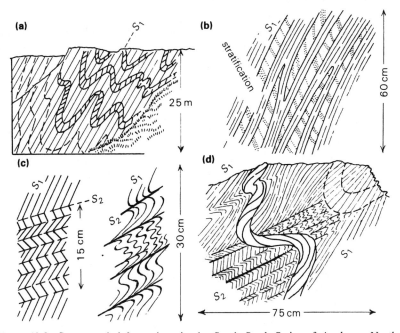

Figure 12.3 Superposed deformations in the South Stack Series of Anglesey, North Wales. (a) First stage folding of greywacke and slate, with the development of the S_1 axial plane cleavage. (b) Sliced slate and greywacke, with original bedding laminae still visible, becoming foliated parallel to S_1, and forming an 'edgewise conglomerate'. (c) Foliated greywacke and slate (S_1), as in (b) above, cut and becoming folded by movement on S_2. (d) Interbedded greywacke with S_1 fracture cleavage, and slate with S_1 schistosity, folded together by second (S_2) movement.

In many parts of the Scottish Highlands, and indeed elsewhere, one may observe linear structures such as rods, parasitic folds, and microfolding in crystalline schists, which cross each other. In some places such structures may not actually intersect, but may curve from one common orientation into another. King and Rast (1955) working in the Dalradian metasediments of Perthshire found that both major and minor folds showed two preferred orientations: one north-east : southwest, the normal Caledonian direction, and the other at right angles to it. In their paper they illustrate examples in which both directions can be seen in the same specimen. In a second paper (King & Rast 1956) they discuss further examples, this time in the Cowal District. Here the first structures to have been formed are isoclinal folds which have two preferred directions of plunge: towards the south-east at about 40°, and towards the south-west at about 5° to 10°. The axial planes of both sets of folds are parallel to each other and to the local schistosity. The authors consider that these folds were contemporaneous and that they form the first structures which were then deformed by later movements, as described by Clough (1897). Another detailed study of small-scale folds in the Dalradian of West Scotland and of Donegal, Ireland, has been made by Professor John and Dr Diane Knill (1958). These authors found that folds which were normally accordant with the major structure were, in some localities, twisted, so that their axes steepen and plunge down the dip of the axial-plane schistosity of the major structure. It is noticeable that the folds which behaved in this discordant manner are dominantly 'detached folds', that is to say, fold hinges which had become separated from their flanks. The sequence – folding, detachment of the fold hinges, and twisting of the fold hinges – is considered by the authors to have resulted from one continuous movement. In the area in Donegal which they investigated, the Drs Knill found that some of the folds and detached fold-hinges lay in a schistose host-rock, of which the schistosity was not in accord with the axial planes of the folds. Such a relationship, they considered, resulted from two successive movements: the first was responsible for the folding, and its tectonic direction would be shown by the orientations of the fold-axes; the direction of the second, oblique to the first, can be deduced from the orientation of the schistosity.

The superposition of large-scale tectonic structures in the Dalradian rocks has been recognised by Weiss and McIntyre (1957). The area discussed lies in the heart of the classic Ballachulish district originally mapped by Bailey (1910, 1922, 1934). The two authors have made a detailed investigation, accompanied by statistical analysis of the orientations of the minor structures in the area, which they have separated into two groups – one early, one late. The first they consider are related to a series of recumbent folds and gliding sheets with a NW–SE tectonic b-direction. The second movement resulted in the deformation of the early structures and folds on NE–SW axes by later folds which have steeply dipping axial planes. It is not known how long a period separated the two movements. The authors' conclusions have been criticised by Bailey (1959) and King and Rast (1959). A reply is given in Weiss and McIntyre (1959).

The formation of domes and basins, or fold culminations and depressions by cross-folding in regions of large-scale simple folding is well known (Moore & Trueman 1939). Even on a small scale one may find such structures; and examples in which the foliation or stratification may form complete steep-sided cylinders or cones, and thus appear on the smooth surface of an exposure as concentric rings have been colloquially referred to as 'eyed folds'.

Analyses of the form of outcrop that one might expect in an area where a pile of flat recumbent folds had been affected by cross-folding have been made by Reynolds and Holmes (1954). Models of recumbently-folded structures were constructed in plasticene, and were then folded across the axes of the first folding; the upper surfaces of the models were then sliced off with a sharp knife to show the internal patterns as they would appear on a horizontal surface of erosion. The models almost duplicate the structure mapped by White and Jahns (1950) in the neighbourhood of Strafford Village, Vermont, USA. Here a group of recumbent folds have been arched upwards at right angles to their fold hinges, so that the fold axes now plunge outwards, away from the region of elevation. The directions and amounts of the fold-plunges were obtained by measuring the orientations of minor folds in the vicinities of the fold-hinges. Between the two groups of folds, the axial-plane schistosity has been gently domed into a cleavage arch; unfortunately linear structures within the arch are not shown. Ramsay (1967, p. 531) shows the outcrop patterns that result when cross-folding has occurred. The outcrop pattern depends upon the relative orientation of the fold hinges and axial planes of the two sets of folds.

The folding of linear structures, rods, small-scale folds and gauffrage of schistosity is generally an indication that more than one phase of movement has been operative in an area. As King and Rast (1955, 1956) have suggested, two sets of folding or crossed lineations may be generated together; nevertheless it is not uncommon to observe that one group of structures is definitely later than the other.

Crossed major and minor structures of different ages are well known in the Moine Series of north-west Scotland, and an excellent summary of studies of such structures has been made by Clifford *et al.* (1957). An account of the results of the investigations, on which this paper was based is given in Sutton (1960). Attention is drawn in these papers to Ramsay's work at Loch Monar (1958a). Here an east–west syncline, B_1, with accordant linear structures, l_1, is refolded by a series of NW–SE folds, B_2, which were also accompanied by linear structures, l_2. The l_1 structures can be traced obliquely up and down over the B_2 folds, and they are deformed where they are crossed by the l_2 structures. The behaviour of the major structures is reflected in that of the small-scale structures. Similar phenomena have been observed by the same author at Glenelg, on the west coast of Scotland (Ramsay 1958b). Here, three sets of folds, $B_1 \, B_2 \, B_3$, have been recognised, and their age relationships can be demonstrated in the field by the manner in which their respective axial-plane schistosities, $S_1 \, S_2 \, S_3$, and their linear structures, $l_1 \, l_2 \, l_3$ deform and offset each other. It is of interest to note that because the later folds are folding rocks which have already been folded, the dips of the limbs of

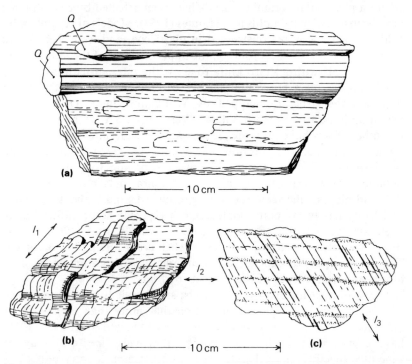

Figure 12.4 (a) Undeformed quartz rods (Q) in siliceous schist, Ben Hutig, A'Mhoine, Sutherland, Scotland; (b), (c) superposed minor structures: rods, small-scale folding and lineations of three generations – l_1, l_2, l_3 Moine Series, Arnisdale, Inverness-shire, Scotland (after Ramsay 1960).

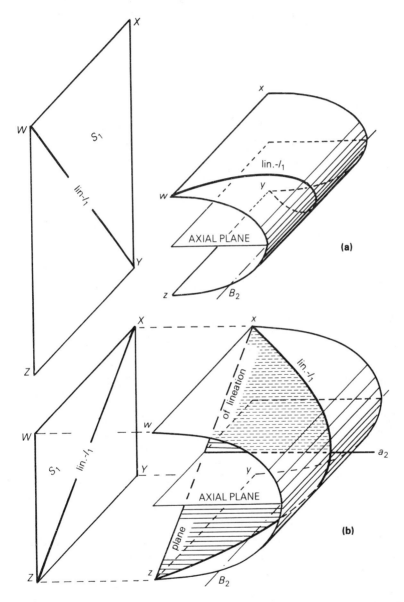

Figure 12.5 (a) The folding of an early lineation l_1 on a surface S_1, by a flexure fold B_2. (b) The folding of an early lineation l_1 on a surface S_1 by a shear-fold B_2 (after Weiss 1959, Ramsay 1960).

the first folds, B_1, exert a control on the plunges of the second folds, B_2, and on the orientations of their associated linear structures, l_2. These last show a marked deflection as they are traced from one side of a first fold to the other. They converge or diverge in their plunge direction, depending on whether the first fold was an antiform or a synform (Clifford *et al.* 1957, Fig. 11).

The deformation of an early lineation, l_1, by a fold, B_2, has been investigated by Sander (1948), Phillips (1954), and Ramsay (1960) (Fig. 12.5). If the surface S_1 which contains the lineation l_1 has been bent by a flexural slip fold, B_2, it has been found that the angle between the folded lineation l_1 and the fold axis B_2 remains constant. On a circular cylindrical surface the lineation would form a helix (Fig. 12.5a).

By unrolling the fold, the original direction of the lineation l_1 as it would appear on a plane surface, can be obtained (Phillips 1954, p. 32, *et seq.*).

If, however, the original surface, which contains the lineation, l_1, has been deformed by shear folding, it cannot be unrolled, as demonstrated by E. Cloos (1947). The fold is not the result of flexure, but has been formed by slip on innumerable parallel cleavage surfaces; and each element on the folded surface has been transported forward parallel to the next. A straight line, on the original surface before it was folded, will therefore be transported forward in a plane while the surface itself is being folded. If the original surface (S_1) is transformed *by slip* into a circular cylinder, the folded lineation on it will form part of an ellipse, not a helix as in the case of flexural folding. The plane of this ellipse will contain the direction of tectonic transport. This direction also lies in the axial plane (S_2) of the second fold: hence, the true direction of transport, a_2, must lie along the line of intersection of these two planes – (i) the axial plane or plane of schistosity (S_2) of the fold B_2, and (ii) the plane containing the folded lineation l_1 (Fig. 12.5b). As can be seen in the figure, the *true direction of transport* (a_2) *in a cleavage fold is not therefore necessarily at right angles to the axis of the fold*, B_2.

In the field one can usually measure the orientation of the plane containing the folded lineation l_1 without much difficulty. The measurement of the orientation of the axial plane of the fold B_2 is not so easy; and it is advisable to check the accuracy of the observation by measuring also the plunge of the axis B_2, which should lie in the axial plane of the fold. The direction a_2 can then be obtained by geometrical construction; or more simply by means of a stereogram. The plotting of the direction of transport of superposed folds by this method is more accurate, and usually yields a more constant tectonic a-direction, than can be obtained from the fold-axes themselves. Ramsay (1960) gives an example of the use of this construction in the elucidation of the tectonics of an area containing complex superposed structures.

13 Minor structures and large-scale tectonics

Throughout the foregoing chapters, I have described the more important varieties of minor structures and their development as individuals or units. Such isolation of individual structures is rare, and generally one finds that more than one group is present. Slight differences in lithology, variations in tectonic style, rates of movement, and the relative ages of impulses during the evolution of the major structure will all influence the types of structures that may be produced and their relationships to each other.

The combinations are legion. For example, in Figure 13.1 we have a fold of strata of varying lithology in which fracture cleavage, axial-plane schistosity and parasitic drag-folds are all present, and are all in accord. Large-scale examples are illustrated in Figure 13.2a, b and c. The first is a simplified and generalised section across an area of folded beds in the Moine Series of the Ross of Mull, Isle of Mull, west Scotland. The synclinal structure of the region is declared by the current bedding in the psammitic beds on the flanks of the fold: the tops face inwards. The metasediments are strongly drag-folded, and the majority of these folds indicate a relative movement of the strata upwards, out of the syncline. Some minor folds have a *régard* in the opposite direction, and probably indicate the inverted limbs of larger folds. Fracture cleavage in these metamorphic rocks is not common; but, where it does occur, its orientation indicates a movement in accord with that shown by the drag-folds.

The style of the folding changes as one traverses the ground from the WNW towards the core of the syncline at Uisken. Here the fold axial planes steepen, the folds themselves become tightly compressed and isoclinal. The pelitic layers in this central section show a clean axial-plane schistosity (Fig. 10.2b) and the more rigid beds are locally boudinaged (Fig. 9.2b). Boudinage also occurs on the eastern flank of the major structure, and here the individual boudins are partially rotated (Fig. 9.2a).

Minor structures and large-scale tectonics 105

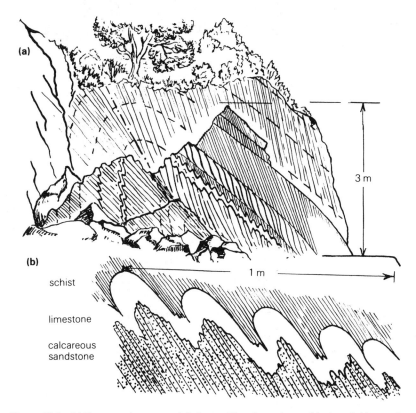

Figure 13.1 (a) Fracture cleavage, axial plane schistosity and parasitic drag-folds together in a single fold, Combe Martin, N. Devon. (b) Details of the parasitic drag-folding.

The sense of rotation agrees with that of the slip of the beds. Near the eastern edge of the section, Clough (1911) had noted felspathic pebbles which were drawn out: the orientation of their longest axes is approximately at right angles to the plunge of the major structure. Several beds of garnet–mica schist are present on both sides of the main syncline. The schistosity is coarsely rippled into microfolds which have a remarkably constant plunge.

The axes of these various minor structures have been plotted on stereograms and, though there are a few anomalies which await further explanation, the majority form a concentrated statistical maximum corresponding to the plunges of the larger-scale folds which can be observed and mapped on the ground. This plunge is between 20° and 25° to the SSW.

The thrust, which cuts the beds west of Uisken, is later than the folding

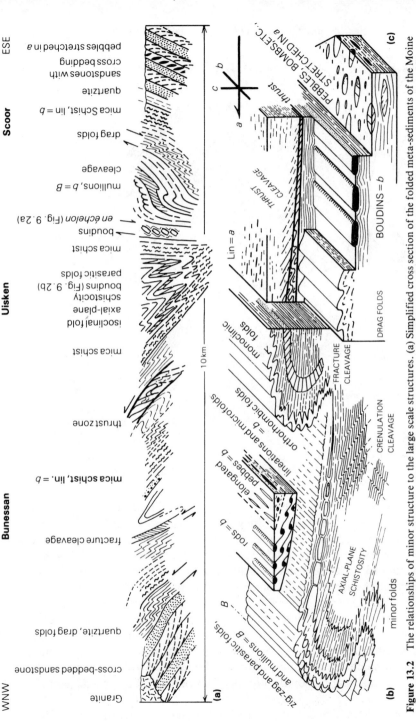

Figure 13.2 The relationships of minor structure to the large scale structures. (a) Simplified cross section of the folded meta-sediments of the Moine Series in the Ross of Mull, Isle of Mull, Scotland; (b) recumbent fold structures, based on the general structure found on A'Mhoine, in North Sutherland, Scotland; (c) an area of flat thrust tectonics, based on the Tintagel Area, Cornwall (after Wilson 1951, Fig. 10).

and metamorphism. Schistose rocks in this zone are strongly sheared, thin siliceous beds are much plicated and fold hinges are, in places, detached. Thicker siliceous beds are folded and sliced by fracture cleavage, and massive formations of quartzite have been cut by subsidiary thrust planes. Here again the minor structures are in general accord, and all indicate the same sense and direction of movement throughout the zone.

In contradistinction to a region in which the deformation has been by folding, two examples where the tectonic styles are dominantly those of horizontal sliding are illustrated in Figure 13.2b and c. The former is an idealised diagram showing the grouping of the different varieties of structures observed in the Moine Series of the A'Mhoine area in Northern Scotland. Here the main movement has been one of horizontal slip combined with overturned and recumbent folding.* An early schistosity parallel to the stratification was produced by this slip, but where folding has taken place an axial-plane schistosity has also developed. Zig-zag minor folds and mullions occur in the hinge zones of the larger folds. Conglomerate pebbles have been flattened in the planes of schistosity and drawn out parallel to the fold axis (i.e. in *b*); they have also been rolled into long rods or spindles, the elongation of which is also parallel to the axes of folding. Segregation quartz rods have also developed in the hinge zones of minor folds. Some of the tectonic units are separated by zones of *décollement* formed of mica schist or siliceous schist. Microfolding forming a ripple lineation in *b* in these schists is common. Throughout the area, the symmetries of the major and minor structures of this tectonic phase are accordant; the *b* axes of the minor structures correspond to the *B* axes of the folds: the structures are homoaxial. The strain-slip cleavage shown really belongs to a later movement phase than the one responsible for the structures discussed above, but it has nevertheless been included in the block diagram.

The third diagram (Fig. 13.2c) is based on the structure seen in the Trebarwith Cliffs of the Tintagel area in north Cornwall (Wilson 1951, Fig. 10, p. 416 and Pl. XXIX). The rocks are of Upper Devonian age, and consist of slates, phyllites and chlorite schists; these last are derived from a group of volcanic rocks. The schistosity and stratification are shown as being horizontal, though in fact they are warped and faulted by later movements. A zone of thrusting which is approximately parallel to the stratification traverses the area.

---

* The folds and minor structures plunge eastwards between 20° and 30°, and are cut off by the Moine Thrust Zone on the west. In my opinion, these structures in the Moine Series are older than, and entirely unrelated to, the Postcambrian movements towards the WNW on the Moine Thrust.

The sense of the movement responsible for the structures is shown by the *regard* of the drag folds, and by the cleavage which occurs within, and in the vicinity of the thrust zone. The *b*-tectonic direction is declared by the axes of drag-folds and by the lengths of boudins, which are found, not only below the thrust zone, but also scattered throughout the rocks of the whole area. Their presence indicates the importance of stretching in the development of the schistosity (Fourmarier 1949). Elongation of lava pillows, bombs, breccia fragments, etc. is strongly marked, and it is parallel to *a* (the direction of movement on the thrust). Their deformation corresponds to that of the *Cyrtospirifer ('Spirifer') verneuili* and other fossils described by Sharpe in 1846. Pre-tectonic spots of metamorphic origin, paratectonic chlorites and acicular tourmaline crystals, and slickenside striations all show this common direction of orientation. In one locality, outside the area illustrated, symmetrical upright folds of 20–30 cm amplitude, whose axes are parallel to the direction of movement *a*, appear in the thrust zone. The lava pillows or bombs, being more rigid than the chlorite schist which encloses them, are in some cases cut by an oblique fracture cleavage, as indicated on the diagram. This results in their having a monoclinic symmetry which is accordant with that of the other structures.

Comparison of the structures of the two areas – A'Mhoine and Tintagel – is of interest. Both are characterised by interstratal gliding, though in A'Mhoine folding has also taken place on a fairly extensive scale. Stretching in *a* is dominant in the Tintagel area, and *b* structures are subordinate. On A'Mhoine and in the Ross of Mull area of Moinian rocks, elongation and lineation in *b* are ubiquitous, while structures in *a* are rare. The rocks of both these latter areas are stratified crystalline schists of biotite–garnet grade; but at Tintagel the metamorphism is predominantly of chlorite grade. Nevertheless, metamorphic grade cannot be considered as controlling tectonic style though it will control the mineralogy, and may give an indication of the physico-chemical conditions under which the deformation took place.

Dominance of stretching features that declare the principal stretching or extension direction (*a*) over structures that declare the intermediate extension direction (*b*) (which may in practice have no extension or even a contraction along it) depends not only upon the relative magnitude of the three principal stresses which determine whether the deformation will be constrictional or flattening (see Flinn 1962) but also upon the orientation of the principal stresses with respect to the rock layering.

When the maximum compression direction is parallel to, or at a low angle to, the layering, folds and other structures that declare the *b*-direction readily form (Fig. 13.2a and b). When the maximum compression is normal to or at a high angle to the layering, folds do not occur and stretching fabrics which declare the principal extension direction a, dominate.

The influence of the relative magnitude of the principal stresses on the structures that develop is well illustrated by the great gravity slides of the Naukluft Mountains of SW Africa described by Korn and Martin (1959). They found that near the base of the complex where deformation was confined and under a heavy load, stretching occurred in a (the direction of movement of the slide). However, in the upper portions of the sliding, folding of the beds under less restricted conditions resulted in the stretching of pebbles parallel to the fold axes, that is, in b.

Linear structures, stretched pebbles, mullions and monoclinic folds transverse to, and parallel to the direction of regional tectonic transport, have been described from the Caledonian orogenic belt in Scandinavia and Scotland. Kvale (1948, 1953) recognised both in the Bergsdalen area of Norway and Lindstrom (1955, 1957, 1958) found similar structures in Northern Sweden. Minor structures at right angles to the main direction of tectonic transport occurred where fold-tectonics were dominant towards the more central portions of the orogen: $B = b$. As one approached the front of the orogen, where the tectonic style became that of flat thrusting, the minor folds and lineations curved around till they were parallel to the direction of movement, that is $B = a$. In the fold zone at the edge of the foreland, structures parallel to the tectonic b once more appeared. Many of the minor structures which Kvale noted as being in a are similar in appearance to those which I have discussed in this paper, and have considered to be b-structures. This similarity is so great, that when Kvale visited the Scottish Highlands he suggested that many of the structures, which he was shown there, were really in a. That is, that the axes of rods, mullions, etc., were parallel to the movements which culminated in the NNW drive of the Moine Thrust (Kvale 1953). He thus supported Anderson's hypothesis that the NW–SE linear structures of the Highlands were parallel to the direction of tectonic transport – in a (Anderson 1948, 1951).

Within the Moine Thrust Zone itself, one finds that in some places the Cambrian and Torridonian beds are over-folded towards the WNW, elsewhere they are sliced by imbricate (*schuppen*) structure which is in accord with the overfolding. These rocks are crushed and locally schisted by the movements, but they are not mylonitised. Mylonite, derived from the Moinian metasediments and the Lewisian Gneisses only occur in the great zone of dislocation; and Christie (1960) observed overturned folds in this mylonite, the axes of which were parallel to the Postcambrian direction of movement, that is ESE–WNW. Johnson (1956) has described orthorhombic 'conjugate folds' and lineations in the same environment; these also have a similar axial orientation. Christie considered that the overturned folds in the mylonite resulted from a lateral movement along the Thrust, and, because the Cambrian quartzites are also lineated in the same axial direction, that the structures resulted from an important lateral movement along the line of the Thrust Zone in Postcambrian times. Since then, Johnson (1960) has made a detailed analysis of the minor structures in a section of the zone of dislocation. He considered that the Postcambrian thrust movements towards the WNW were the last to take place, and that the Moine Thrust as a whole has had a *long and discontinuous* movement history; and that the transverse structures which plunge to the SE in the thrust zone 'bear no relation to the generally accepted pattern of thrust movements'. The case for the development of folds about a in the Scottish Highlands has received little support in recent publications.

In the meanwhile, the problem has been carried into the Caledonides of Swedish Lapland. Maurits Lindstrom (1955, 1957, 1958) has described many varieties of linear and monoclinic structures whose b-axes are parallel to the regional direction of tectonic transport. In his 1957 paper, he compared the folding observed, in this district of typically thrust tectonics, with that found in salt diapirs where there is an all round restriction to movement save in one direction. He referred to Balk (1949) and Ahlborn and Richter-Bernbugg (1955) who have described and illustrated remarkable fold patterns in salt plugs, the axes of which are vertical parallel to the movement. It is of interest also to recall that Eskola (1951) noted folds with vertical axes in the granite gneiss dome of Sumeria, and that Pitcher and Read (1959) found, on the margin of the Main Donegal Granite, mullions which were oriented parallel to the probable direction of flow of the granite. In the same year that Lindstrom's paper appeared, Turner (1957) marshalling heavy batteries of experimental and other evidence based on *Gefuge* studies, severely criticised those workers who assumed that monoclinic minor structures could arise from movements perpendicular to their planes of symmetry. Lindstrom in 1957 had already noted that though some of his monoclinic structures had a *régard* in one direction, an equal number faced the other way. In 1958 he discussed Turner's criticisms, and considered that, in a zone of deep-seated thrust tectonics, unless the small-scale monoclinic structures showed a constant sense of *régard over a large area*, there was no evidence for assuming any tectonic transport at right angles to their b-axes; and that folding under restricted conditions around the regional a-axis could take place. The debate continues . . .

14 Conclusions

Structural geology is a branch of our science which is becoming more and more important; and, like any other scientific subject which is actively developing, is becoming more and more involved. Rocks are now being seen in different perspectives from those known 30 years ago, and new techniques for their investigation are rapidly being evolved. But, no matter what geological problem we are endeavouring to solve, it is in the field that we must collect our data; and it is there again that the final correctness of our solution must be demonstrated. It is therefore for the field geologist that I have written. I have endeavoured to describe the small-scale structures which he is liable to meet, and to discuss their uses and limitations in elucidating the geometry and tectonic history of an area.

Readers who are specialists in structural geology and *Gefugekunde* may accuse me of over-simplification of my subject, and of sins of omission − perhaps even of commission! But I have not written for specialists, rather I have tried to explain to the non-specialist the data for which the specialist searches, and to outline the general principles by which he reaches his conclusions.

Much still remains to be done, and as our knowledge and experience grow, we may find complications in what we originally saw as simple problems. The mechanics of some of the structures I have discussed are not really clear, and some are frankly doubtful. For example, we have not yet got a definite explanation for the production of fracture cleavage; in fact there are some geologists who doubt if such a structure *per se* does exist at all. The direction of movement responsible for shear folds was, until a short time ago, considered to be normal to fold axes. Now, as a result of Ramsay's work, it can be shown that the symmetry of the folds may not be accordant with the symmetry of the movement responsible for them. The problem as to whether or not folds or other monoclinic structures can actually develop in *a*, parallel to the direction of movement, is one which, in some circles, it may be positively dangerous to discuss!

Although further understanding of geological structures, their formation and what they mean is still needed, we must not lose sight of the ultimate aim of field geology. The tendency to concentrate on the structure, the whole structure and nothing but the structure, to the neglect of stratigraphical mapping is unfortunately very strong, and must be guarded against. It must be appreciated that the geometry of the structures observed, and the kinematics or dynamic systems which may be deduced from it, are but instrumental aids to the final elucidation of the geological history.

The accumulation of small-scale structural data is a painstaking but fascinating business. It involves plotting not only dips and strikes of stratification, but also of schistosity, together with the plunges of linear structures, and the collection of oriented specimens for laboratory study. It is easy to be carried away by enthusiasm for the subject while in the field and to miss other equally important but less exciting geological observations (*peccavi*). This means that the worker must submit himself to a strict discipline, because without corresponding attention to stratigraphy and its kindred subjects, a tectonic or historical synthesis of a region may be incorrect, or at least incomplete.

The converse is also true; because without appreciation of the movements which the rocks have suffered, the stratigraphy itself may be incorrect or undecipherable. Charles Lapworth's advice to a student who was in difficulties in the field – 'Map it, my boy, and it will all come out' – is as true today as it was over 70 years ago; but in addition should be added the dictum enunciated by Ernst Cloos: '. . . every structure in a rock is significant, none is unimportant, even if at first it may seem irrelevant'.

Bibliography

Adams, F. D. 1912. An experimental contribution to the question of the depth of the zone of flow in the Earth's crust. *J. Geol.* **20**, 97.
Adams, F. D. and J. A. Bancroft 1917. On the amount of internal friction developed in rocks during deformation, and on the relative plasticity of different kinds of rocks. *J. Geol.* **25**, 597.
Aderca, B. M. 1960. Schistosité de dislocation. *Ann. Soc. Géol. de Belgique* **83**, 101.
Ahlborn, O. and G. Richter-Bernburg 1955. Excursion zum Salzstock von Benthe (Hannover) mit Befahrung der Kaliwerk Ronnenberg und Hanza. *Z. D. Geol. Gesellsch.* **105** (1953), 855.
Anderson, E. M. 1948. On lineation and petrofabric structure and the shearing movement by which they have been produced. *Q. J. Geol Soc. Lond.* **104**, 99.
Anderson, E. M. 1951. The dynamics of faulting, 2nd edn. Edinburgh: Oliver & Boyd.

Baer, A. 1956. La schistosité et sa répartition. *Geol. Rundsch.* **45**, 234.
Bailey, E. B. 1910. Recumbent folds in the schists of the Scottish Highlands. *Q. J. Geol Soc. Lond.* **66**, 586.
Bailey, E. B. 1922. The structure of the south-west Highlands of Scotland. *Q. J. Geol Soc. Lond.* **78**, 82.
Bailey, E. B. 1930. New light on sedimentation and tectonics. *Geol Mag.* **67**, 77.
Bailey, E. B. 1934. West Highland tectonics: Loch Leven to Glen Roy. *Q. J. Geol Soc. Lond.* **90**, 462.
Bailey, E. B. 1935. *Tectonic essays.* Oxford: Clarendon Press.
Bailey, E. B. 1959. Structural geometry of Dalradian rocks at Loch Leven: a discussion. *J. Geol.* **67**, 246.
Bailey, E. B. and W. J. McCallien 1937. Perthshire tectonics: Schiehallion to Glen Lyon. *Trans R. Soc. Edinb.* **59**, 79.
Baily, B., G. Borradaile and C. McA. Powell 1977. *Atlas of rock cleavage.* Tasmania University Press.
Balk, R. 1937. *Structural behaviour of igneous rocks.* Geol Soc. Am. Mem. no. 5.
Balk, R. 1949. Structure of Grand Saline Salt Dome, Van Zandt County, Texas. *Bull. Am. Assn Petrol. Geol.* **33**, 1791.
Balk, R. and T. F. W. Barth 1936. Structural and petrologic studies in Duchess County, New York. *Bull. Geol Soc. Am.* **47**, 685.
Barrow, G., J. S. Grant-Wilson and E. H. Cunningham-Craig 1905. *The geology of the country around Blair Atholl, Pitlochry, etc.* Mem. geol. surv. Scotland, sheet 55, Pl. III.
Becke, F. 1924. Struktur und Kluftung. *Fortschr. Min. Krist. Petr.* **9**, 185.
Becker, G. F. 1882. *Geology of the Comstock Lode.* US Geol. Surv. monogr. no. 3.

Becker, G. F. 1893. Finite homogeneous strain, flow and rupture of rocks. *Bull. Geol Soc. Am.* **4**, 13.
Behre, C. H., Jr 1933. *Slate in Pennsylvania.* Penn. Geol Surv., 4th ser., bull. M 16.
Billings, M. P. 1942. *Structural geology.* Englewood Cliffs, NJ: Prentice-Hall.
Bloomer, R. O. and H. J. Werner 1955. Geology of the Blue Ridge region in Central Virginia. *Bull. Geol Soc. Am.* **66**, 579, 605.
Blyth, F. G. H. 1950. The sheared porphyry dykes of South Galloway. *Q. J. Geol Soc.* **105**, 393.
Bonney, T. G. 1886. Anniversary address of the president. *Q. J. Geol Soc. Lond.* **42**, 95, 98–9.
Bonte, A. 1953. *Introduction à la lecture des cartes géologiques*, 2nd edn, 181. Paris.
Born, A. 1929. Ueber Druckschieferung im varistischen Gebirgskörper. *Fortschr. d. Geol. u. Pal.* **VII**, 329.
Bosworth, T. O. 1910. Metamorphism around the Ross of Mull Granite. *Q. J. Geol Soc. Lond.* **56**, 376, 381.
Briggs, H. 1927. An attempt at the rationale of faulting and subsidence. *Trans Inst. Min. Engrs Lond.* **73**, 465.
Broughton, J. G. 1946. An example of the development of cleavages. *J. Geol.* **54**, 1.
Bull, A. J. 1922. In Evans, J. W. Excursion to Combe Martin. *Proc. Geol Assoc.* **33**, 228–34, Pl. 7A.

Campbell, J. D. 1951. Some aspects of rock folding by shearing deformation. *Am. J. Sci.* **249**, 625.
Campbell, J. D. 1958. *En échelon* folding. *Econ. Geol.* **53**, 448.
Carey, S. W. 1954. The Rheid concept. *Geol Soc. Aust. J.* **1**, 67.
Christie, J. M. 1960. Mylonitic rocks of the Moine Thrust-Zone in the Assynt region, north-west Scotland. *Trans Geol Soc. Edinb.* **18**, 79.
Clarke, R. H. and D. B. McIntyre 1951. The use of the terms pitch and plunge. *Am. J. Sci.* **249**. 591.
Clifford, P. 1960. The geological structure of the Loch Luichart area, Ross-shire. *Q. J. Geol Soc. Lond.* **115** (1959), 365.
Clifford, P., M. J. Fleuty, J. G. Ramsay, J. Sutton and J. Watson 1957. The development of lineation in complex fold systems. *Geol Mag.* **94**, 1.
Cloos, E. 1932. Feather joints as indicators of the direction of movements on faults, thrusts, joints and magmatic contacts. *Proc. Nat. Acad. Sci. Wash.* **18**, 387.
Cloos, E. 1937. The application of recent structural methods to the interpretation of the crystalline rocks of Maryland. *Maryland Geol Surv.* **13**, 29.
Cloos, E. 1942. Distortion of stratigraphic thicknesses due to folding. *Proc. Nat. Acad. Sci. Wash.* **28**, 401.
Cloos, E. 1943. Methods of measuring changes of stratigraphic thicknesses due to flowage and folding. *Trans Am. Geophys. U. 24th Ann. Mtg.* part 1, 273.
Cloos, E. 1946. *Lineation.* Geol Soc. Am. mem. no. 18.

Cloos, E. 1947a. Boudinage. *Trans Am. Geophys. U.* **28**, 626.
Cloos, E. 1947b. Oolite deformation in the South Mountain Fold, Maryland. *Bull. Geol Soc. Am.* **58**, 843.
Cloos, E. 1955. Experimental analysis of fracture patterns. *Bull. Geol Soc. Am.* **66**, 241.
Cloos, E. and A. Heitanen 1941. Geology of the 'Martic Overthrust' and the Glenarm Series in Pennsylvania and Maryland. Geol Soc. Am. special paper no. 35.
Cloos, H. and H. Martin 1932. Der Gang einer Falte. *Forschr. Geol. Palaeont.* **11**, no. 33, Decke Fetschr., 74.
Clough, C. T. 1897. In *The geology of Cowal*, W. Gunn, C. T. Clough & J. B. Hill. Mem. Geol Surv. Scotland.
Clough, C. T. 1911. In *The geology of Colonsay and Oronsay, with parts of the Ross of Mull*, E. H. Cunningham-Graig, W. B. Wright & E. B. Bailey. Mem. Geol Surv. Scotland, sheet 35.
Cobbold, P. R., J. W. Cosgrove and J. M. Summers 1971. The development of internal structures in deformed anisotropic rocks. *Tectonophysics* **12**, 23–53.
Coe, K. 1959. Boudinage structure in West Cork, Ireland. *Geol Mag.* **96**, 191.
Colette, B. J. 1958. On the origin of schistosity, I and II. *Proc. Kon. Ned. Akad. Wet A'dam* **B 61**, 121.
Colette, B. J. 1959. On helicitic structures and the occurrence of elongate crystals in the direction of the axis of a fold. *Proc. Kon. Ned. Akad. Wet. A'dam* **B 62**, 161, 170.
Collomb, P. 1960. La lineation dans les roches. *Bull. trim. S. I. G. du B. R. G. M. (Paris) 12th Ann* (48), 1.
Corin, F. 1932. A propos du boudinage en Ardenne. *Bull. Soc. Belge de Geol., Pal. et Hydrol.* **42**, 101.
Cosgrove, J. W. 1976. The formation of crenulation cleavage. *J. Geol Soc.* **132**, 155–78.
Cosgrove, J. W. 1980. The tectonic implications of some small-scale structures in the Mona Complex of Holy Isle, North Wales. *J. Struct. Geol.* **2** (4), 383–96.

Dale, T. N. 1896. Structural details in the Green Mountain Region and in eastern New York. US Geol Surv., 16th ann. rep. (1894–5) part I, 543.
Dale, T. N. 1899. *The slate belt of eastern New York and western Vermont.* US Geol Surv., 19th ann. rep. (1897–8) part III; 153, Pl. XXIX, E, L.
Darwin, C. 1846. *Geological observations in South America.* London: Smith, Elder & Co.
de la Beche, H. T. 1853. *The geological observer*, 2nd edn. London: Longman.
Derry, D. R. 1939. Some examples of detailed structure in early Pre-Cambrian rocks of Canada. *Q. J. Geol Soc. Lond.* **95**, 109.
Dewey, J. F. 1965. Nature and origin of kink bands. *Tectonophysics* **1**, 459–94.
Dieterich, J. H. and N. L. Carter 1969. Stress history of folding. *Am. J. Sci.* **267**, 129–54.

Eskola, P. 1951. Around Pitkäranta. *Ann. Acad. Sci. Fennicae* A, III Geol. Geogr. **27**, 18–21.

Fallot, P. 1957. *La tour de Babel*. Bologna: Scientia.
Fath, A. E. 1920. *The origin of the faults, anticlines and buried 'granite ridge' of the northern part of the Mid Continental oil and gas field*. US Geol Surv. prof. paper 128, 75.
Fermor, L. L. 1909. *The manganese-ore deposits of India*. Mem. Geol Surv. India no. 37, 935.
Fermor, L. L. 1924. The pitch of rock folds. *Econ. Geol.* **19**, 559.
Flett, J. S. 1912. In *The geology of Ben Wyvis, Carn Chuinneag, Inchbae and surrounding country*, B. N. Peach *et al.* Mem. Geol Surv. Scotland, sheet no. 93.
Flinn, D. 1952. A tectonic analysis of the Muness Phyllite block of Unst and Uyea, Shetland. *Geol Mag.* **89**, 263.
Flinn, D. 1956. On the deformation of the Funzie conglomerate, Fetlar, Shetland. *J. Geol.* **64**, 480.
Flinn, D. 1961. On deformation at thrust planes in Shetland and the Jotunheim area of Norway. *Geol Mag.* **98**, 245.
Flinn, D. 1962. On folding during three-dimensional progressive deformation. *Q. J. Geol Soc.* **118**, 385–433.
Fourmarier, P. 1923. De l'importance de la charge dans le développement du clivage schisteux. *Bull. Acad. R. Belg. Cl. Sci.* **5** (9), 454.
Fourmarier, P. 1932. Observations sur le développement de la schistosité dans les séries plissées. *Bull. Acad. R. Belg. Cl. Sci.* **5** (18), 1049.
Fourmarier, P. 1948. Schistosité regionale et schistosité locale. *Arch. Sci. Phys. Nat. Genève, 153rd Ann.* **1**, 188.
Fourmarier, P. 1949a. *Principes de géologie*, 3rd edn, I and II. Paris-Liège: Masson Liège, H. Vaillant-Carmaune.
Fourmarier, P. 1949b. Etirement des roches et la schistosité. *Bull. Soc. Géol. France* **5** (19), 569.
Fourmarier, P. 1949c. Observations sur le comportement de la schistosité dans les Alpes. *Ann. Hébert et Haug* (Univ. Paris) **7**, 171.
Fourmarier, P. 1951. Schistosité, foliation et microplissement. *Arch. Sci. Phys. Nat. Genève* **4**, 7.
Fourmarier, P. 1952a. Aperçu sur les déformations intimes des roches en terrains plissées. *Ann. Soc. Géol. Belg.* **75**, B181.
Fourmarier, P. 1952b. Microplissement et plis minuscules. *Ann. Soc. Géol. Belg.* **76**, B81.
Fourmarier, P. 1953a. Schistosité et phénomènes connexes dans les séries plissées. *Int. geol Cong. 19th Algeria, C. R. sec. 3, pt 3, 117.*
Fourmarier, P. 1953b. Schistosité et grande tectonique. *Ann. Soc. Géol. Belg.* **76**, B275, B286–7.
Fourmarier, P. 1959. Le granite et les déformations mineures des roches. *Mem. Acad. R. Belg. Cl. Sci.* **31**.
Fuchs, Sir V. and Sir E. Hillary 1958. *The crossing of Antarctica, 1955–1958*. Pl. 4, 20. London.

Gilbert, G. K. 1928. *Studies of Basin Range structure*. US Geol Surv. prof. paper 153, 450.
Gindy, A. R. 1953. The plutonic history of the district around Trewenagh Bay, Donegal. *Q. J. Geol Soc. Lond.* **108** (1952), 377, Fig. 3B.
Goguel, J. 1945. Sur l'origine mécanique de la schistosité. *Bull. Soc. Géol. France* **5** (15), 509.
Goguel, J. 1948. Introduction à l'étude mécanique des déformations de l'écorce terrestre. *Mem. Carte Géol. France*, 2nd edn, 7.
Goguel, J. 1952. *Traité de tectonique*. Paris: Masson.
Goguel, J. 1953. Importance des facteurs physico-chimiques dans la déformation des roches. *Int. Geol. Cong. 19th Algeria, C. R.* sec. 3, pt 3, 133, 153.
Gonzalez-Bonorino, F. 1960. The mechanical factor in the formation of schistosity. *Int. Geol. Cong. 21st Norden, C. R.* pt 18, 303, Fig. 4b.
Gregory, H. E. 1914. The Rodadero (Cuzco, Peru). A fault-plane of unusual aspect. *Am. J. Sci.* **4** (37), 289.
Green, J. F. N. 1917. The age of the chief intrusions of the Lake District. *Proc. Geol Assn Lond.* **28**, 1.
Green, J. F. N. 1924. The structure of the Bowmore-Portaskaig district of Islay. *Q. J. Geol Soc.* **80**, 72.
Greenly, E. 1919. *The geology of Anglesey*, I and II. Mem. Geol Surv. GB.
Griggs, D. 1936. Deformation of rocks under high confining pressures. *J. Geol.* **44**, 541.
Griggs, D. 1938. Deformation of single calcite crystals under high confining pressures. *Am. Mineral.* **23**, 28.
Griggs, D. 1939. Creep in rocks. *J. Geol.* **47**, 225.
Griggs, D. 1940. Experimental flow of rocks under conditions favouring recrystallization. *Bull. Geol Soc. Am.* **51**, 1001.
Griggs, D. and J. Handlin 1960. *Rock deformation (a symposium)*. Geol Soc. Am. mem. no. 79, 335–8, pls 6–10.

Harker, A. 1885. On slaty cleavage and allied rock structures. Brit. Assn Adv. Sci. rep. (1885–6), 813.
Harker, A. 1932. *Metamorphism*. London: Methuen.
Hartley, J. J. 1925. The succession and structure of the Borrowdale Volcanic Series as developed in the area lying between the lakes of Grasmere, Windermere and Coniston. *Proc. Geol Assn Lond.* **36**, 20–3, 203, Pl. 16B.
Haug, E. 1927. *Traité de géologie*, I. Paris: Libraire Armand Colin.
Haughton, S. 1856. On slaty cleavage and the distortion of fossils. *Phil Mag.* **4** (12), 409.
Heim, A. 1878. *Untersuchungen über den Mechanismus der Gebirgsbildung*, I, II and atlas. Basel: Benno Schwabe.
Heim, A. 1919 and 1921. *Geologie der Schweiz*, I and II. Leipzig: Chr. Herm. Tauchnitz.
Hills, E. S. 1940 and 1953. *Outlines of structural geology*, 1st and 2nd edns. London: Methuen.
Hills, E. S. 1945. Examples of the interpretation of folding. *J. Geol.* **53**, 47.

Hitchcock, E., C. H. Hitchcock and A. D. Hager 1861. *Report on the geology of Vermont*, I and II. Claremont, NH.
Hoeppener, R. 1956. Zum Problem der Bruchbildung, Schieferung und Faltung. *Geol. Rundsch.* **45**, 247, 268–70.
Holmes, A. 1928. *Nomenclature of petrology*. London: Thomas Murby.
Holmquist, P. J. 1931. On the relations of the 'Boudinage Structure'. *Geol. fören Stockholm Forh.* **53**, 193.
Hossain, K. M. 1979. Determination of strain from stretched belemnites. *Tectonophysics* **60**, 279–88.
Hubbert, M. K. 1937. Theory of scale models as applied to geologic structures. *Bull. Geol Soc. Am.* **48**, 1459.
Hubbert, M. K. 1951. Mechanical basis for certain familiar geologic structures. *Bull. Geol Soc. Am.* **62**, 355.
Hull, E., G. H. Kinahan and N. Nolan 1891. *Northwest and central Donegal*. Mem. Geol Surv. Ireland sheets 3, 4, etc.

Jannetaz, E. 1884. Mémoire sur les clivages des roches (schistosité, longrain) et sur leur reproduction. *Bull. Soc. Géol. France* **3** (12), 211, 225.
Johnson, M. R. W. 1956. Conjugate fold systems in the Moine Thrust zone in the Lochcarron and Coulin Forest areas of Wester Ross. *Geol Mag.* **93**, 345.
Johnson, M. R. W. 1957. The tectonic phenomena associated with the post-Cambrian thrust movements at Coulin, Wester Ross. *Q. J. Geol Soc. Lond.* **113**, 241.
Johnson, M. R. W. 1960. Linear structures in the Moine Thrust belt of NW Scotland and their relation to the Caledonian orogenesis. *Int. Geol Cong. 21st Norden, C. R.* pt 19, 89.
Jukes, J. Beete 1842. *Excursions in and about Newfoundland*, II, 325–6. London: John Murray.

Kanungo, D. 1956. *The structural geology of the Torridonian, Lewisian and Moinian rocks of the area between Plockton and Kyle of Lochalsh in Wester Ross, Scotland*. Unpublished PhD Thesis, University of London.
Karman, von Th. 1911. Festigkeitversuche unter allseitigem Druck. *Z. Ver. dtsch. Ing.* **55**, 1749.
Kautsky, G. 1953. Der geologische Bau des Sulitelma-Salojauregebietes in den Nordskandinavischen Kaledoniden. *Sverig. Geol. Unders. Avh.* C, no. 528 (1952), 131, Fig. 61.
Kenny, J. P. L. 1936. Golden Stairs Mine, Greensborough. *Geol Surv. Victoria, Aust.* **5**, 222.
Kienow, S. 1942. Grundzüge einer Theorie der Faltungs und Schieferungsvorgänge. *Fortschr. Geol. Paläont.* **14** (46), 126.
King, B. C. and N. Rast 1955. Tectonic styles in the Dalradians and Moines of parts of the Central Highlands of Scotland. *Proc. Geol Assn Lond.* **66**, 243.
King, B. C. and N. Rast 1956. The small-scale structures of south-eastern Cowal, Argyllshire. *Geol Mag.* **93**, 185.
King, B. C. and N. Rast 1959. Structural geometry of Dalradian rocks at Loch Leven, Scottish Highlands: a discussion. *J. Geol.* **67**, 244.

Knill, J. L. 1960. A classification of cleavages, with special references to the Craignish district of the Scottish Highlands. *Int. Geol. Cong. 21st Norden, C. R.* pt 18, 317.
Knill, J. C. and D. C. Knill 1958. Some discordant fold structures from the Dalradian of Craignish, Argyll, and Rosguill, Co. Donegal. *Geol Mag.* **95**, 497.
Knopf, A. 1941. *Petrology.* Geol Soc. Am., 50th Anniv. Vol., 335.
Knopf, E. B. and E. Ingerson 1938. *Structural petrology.* Geol Soc. Am. mem. no. 6.
Korn, H. and H. Martin 1959. Gravity tectonics in the Naukluft Mountains of south-west Africa. *Bull. Geol Soc. Am.* **70**, 1047, Fig. 15, 1070.
Kranck, E. H. 1960. On lineation in gneisses and schists. *Bull. Comm. Géol. Finlande* no. 188, 11.
Krige, L. J. 1916. Petrographische Untersuchungen im Val Piora und Umgebung. *Eclog. Geol. Helvet.* **14**, 519.
Kuenen, Ph. H. and L. U. de Sitter 1938. Experimental investigation into the mechanism of folding. *Leidsche Geol. Med.* **10**, 217, Figs 14 and 15.
Kvale, A. 1948. Petrologic and structural studies in the Bergsdalen Quadrangle, Western Norway, pt II. *Bergens Mus. Arbok* no. 1 (1946–7).
Kvale, A. 1953. Linear structures and their relation to movement in the Caledonides of Scandinavia and Scotland. *Q. J. Geol Soc. Lond.* **109**, 51.

Lamont, A. 1940. First use of current bedding to determine the orientation of strata. *Nature, Lond.* **145**, 1016.
Lapworth, C. 1893. *Presidential address to Section C (Geology).* Brit. Assn Adv. Sci. rep. 1892, 695.
Laugel, A. 1855. Du clivage des roches. *Bull. Soc. Géol. France* **2** (12), 268, 366.
Lawson, A. C. 1913. *The Archean geology of Rainy Lake restudied.* Geol. surv. Canada mem., no. 40; 63, Fig. 1.
Leith, A. 1931. The application of mechanical structural principles in the Western Alps. *J. Geol.* **39**, 625.
Leith, C. K. 1905. *Rock cleavage.* US Geol Surv., bull. no. 239.
Leith, C. K. 1923. *Structural geology.* New York: Henry Holt.
Lewis, H. P. 1946. Bedding-faults and related minor structures in the Upper Valentian rocks near Aberystwyth. *Geol Mag.* **83**, 153.
Lillie, A. R. 1961. Folds and faults in the New Zealand Alps and their tectonic significance. *Proc. R. Soc. NZ, 9th Sci. Cong. Rep.* **89**, 57.
Lindström, M. 1955. Structural geology of a small area in the Caledonides of Arctic Sweden. *Lunds Univ. Arsskrift. NF* Adv. 2 51, no. 15.
Lindström, M. 1957. Tectonics of the area between Mt Keron and Lake Allesjaure in the Caledonides of Swedish Lapland. *Lunds Univ. Arsskrift. NF* Adv. 2 53, no. 11.
Lindström, M. 1958. Tectonic transports in three small areas in the Caledonides of Swedish Lapland. *Lunds Univ. Arsskrift. NF* Adv. 2 54, no. 3; Pl. III, 2.
Lohest, M., X. Stanier and P. Fourmarier 1909. C. R. de la session extraordinaire de la Soc. Géol. de Belgique tenue à Eupen et à Bastogne, 29 août au 3 sept. 1908. *Ann. Soc. Géol de Belgique* **35**, 351.

Lovering, T. S. 1928. The fracturing of incompetent beds. *J. Geol.* **36**, 709.

Macculloch, J. 1819. *A description of the Western Islands of Scotland . . . comprising an account of their geological structure . . . etc.* I–III. London: Hurst, Robinson & Co.

McIntyre, D. B. 1950. Note on lineation, boudinage and recumbent folds in the Struan Flags (Moine) near Dalnacardoch, Perthshire. *Geol Mag.* **87**, 427.

McIntyre, D. B. 1951. The tectonics of the area between Grantown and Tomintoul (Mid-Strathspey). *Q. J. Geol Soc. Lond.* **107**, 1.

McLachlan, G. R. 1953. The bearing of rolled garnets on the concept of b-lineation in Moine rocks. *Geol Mag.* **90**, 172.

Margerie, E. de and A. Heim 1888. *Les dislocations de l'écorce terrestre.* Zurich: J. Warster.

Mead, W. J. 1940. Folding, rock flowage and foliate structures. *J. Geol.* **48**, 1007.

Melton, F. A. 1929. A reconnaissance of the joint-systems in the Ouachita Mountains and central plains of Oklahoma. *J. Geol.* **37**, 729.

Mendelsohn, F. 1959. The structure of the Roan Antelope Deposit. *Trans Inst. Min. Metall. Lond.* **68** (1958–9), 229.

Metz, K. 1957. *Lehrbuch der Tektonischen Geologie*, 43. Stuttgart: Ferdinand Enke.

Moore, L. R. and A. E. Trueman 1939. The structure of the Bristol and Somerset coalfields. *Proc. Geol Assn Lond.* **50**, 46.

Moret, L. 1947. *Précis de géologie*, 366. Paris.

Muff, H. B. 1909. In *The geology of the seaboard of Mid-Argyll*, B. N. Peach, H. Kynaston & H. B. Muff (Maufe), 14–17. Mem. Geol Surv. Scotland, sheet no. 36.

Nevin, C. M. 1936 and 1949. *Principles of structural geology*, 2nd edn (141, Fig. 88) and 4th edn. New York: Wiley.

Nieuwenkamp, W. 1928. Measurements on slickensides on planes of stratification in folded regions. *Proc. Kon. Ned. Akad. Wet. A'dam.* **31**, 255.

Otley, J. 1820. *Kirkby Lonsdale Magazine*, quoted in Harker 1885.

Otley, J. 1823. *Concise description of the English Lakes . . .* Keswick.

Oulianoff, N. 1958. Le métamorphisme des roches dans ses rapports avec les mouvements tectoniques. *Bull. Lab. géol. minér. Musée géol. Lausanne*, no. 23.

Parkinson, J. 1903. Geology of the Tintagel and Davidstow district, northern Cornwall. *Q. J. Geol Soc. Lond.* **59**, 408, 410.

Patterson, M. S. and L. E. Weiss 1968. Folding and boudinage of quartz-rich layers in experimentally deformed rocks. *Bull. Geol Soc. Am.* **79**, 795–812.

Peach, B. N. and J. Horne 1907. *The geological structure of the North-West Highlands of Scotland.* Mem. Geol Surv. Scotland.

Peach, B. N., H. Kynaston and H. B. Muff 1909. *The geology of the seaboard of mid-Argyll (explanation of sheet 36).* Mem. Geol Surv. UK, 1–121.

Phillips, F. C. 1937. A fabric study of some Moine Schists and associated rocks. *Q. J. Geol Soc. Lond.* **93**, 581.
Phillips, F. C. 1954. *The use of stereographic projection in structural geology.* London: Edward Arnold.
Phillips, J. 1857. *Report on cleavage and foliation in rocks, and on the theoretical explanations of these phenomena.* Brit. Assn Adv. Sci. rep. no. 1856, 269.
Pilger, A. and W. Schmidt 1957a. Definition des Begriffes 'Mullion-Struktur' (mullion structure). *Neues Jb. Geol. Paläontol. Mh.* 1957, 24.
Pilger, A. and W. Schmidt 1957b. Mullion-Strukturen in der Nord-Eifel. *Abh. hess. Landsamt. Bodenforsch.* **20**.
Pitcher, W. S. and H. H. Read *et al.* 1959. The Main Donegal Granite. *Q. J. Geol Soc. Lond.* **114** (1958), 259, Pl. XI.
Pitcher, W. S. and H. H. Read 1960. The aureole of the Main Donegal Granite. *Q. J. Geol Soc. Lond.* **116**, 1.
Price, N. J. 1966. *Fault and joint development in brittle and semi-brittle rocks*, 1–176. Oxford: Pergamon Press.
Price, N. J. 1975. Rates of deformation. *J. Geol Soc. Lond.* **131**, 553–75.
Pumpelly, R., J. E. Wolff and T. N. Dale 1894. *Geology of the Green Mountains in Massachusetts.* US Geol Surv. monogr. 23, 158.

Quirke, T. T. 1923. Boudinage, an unusual structural phenomenon. *Bull. Geol Soc. Am.* **34**, 650–2.

Ragan, D. M. 1967. Planar and layered structures in glacial ice. *J. Glaciol.* **6**, 565–7.
Ramberg, H. 1955. Natural and experimental boudinage and pinch-and-swell structures. *J. Geol.* **63**, 512.
Ramsay, A. C. 1881. *The geology of North Wales.* Mem. Geol Surv. GB 3.
Ramsay, J. G. 1958a. Superimposed folding at Loch Monar, Inverness-shire and Ross-shire. *Q. J. Geol Soc. Lond.* **113** (1957), 271.
Ramsay, J. G. 1958b. Moine-Lewisian relations at Glenelg, Inverness-shire. *Q. J. Geol Soc. Lond.* **113** (1957), 487.
Ramsay, J. G. 1960. The deformation of early linear structures in areas of repeated folding. *J. Geol.* **68**, 75.
Ramsay, J. G. 1967. *Folding and fracturing of rocks.* New York: McGraw-Hill.
Ramsay, J. G. and R. H. Graham 1970. Strain variations in shear belts. *Can. J. Earth Sci.* **7**, 786–813.
Rast, N. 1956. The origin and significance of boudinage. *Geol Mag.* **93**, 401, Figs 3a and 3b.
Read, H. H. 1934. On the segregation of quartz–chlorite–pyrite masses in Shetland igneous rocks during dislocation–metamorphism, with a note on an occurrence of boudinage-structure. *Proc. Liverpool Geol Soc.* **16** (1933–4), 128.
Read, H. H. 1948. A commentary on place in plutonism. *Q. J. Geol Soc. Lond.* **104**, 155, 185.
Read, H. H. 1949a. *Geology: an introduction to Earth history*, 125. Oxford: Oxford University Press.

Read, H. H. 1949b. A contemplation of time in plutonism. *Q. J. Geol Soc. Lond.* **105**, 101.
Read, H. H. 1957. *The granite controversy*, 279. London: Thomas Murby.
Read, H. H. 1958. Stratigraphy in metamorphism. *Proc. Geol Assn Lond.* **69**, 83.
Read, H. H. and J. Phemister 1926. The geology of Strath Oykell and Lower Loch Shin. Mem. Geol Surv. Scotland, sheet 102, 121.
Reynolds, D. L. and A. Holmes 1954. The superposition of Caledonoid folds on an older fold-system in the Dalradians of Malin Head, Co. Donegal. *Geol Mag.* **91**, 417.
Richey, J. E. and W. Q. Kennedy 1939. The Moine and Sub-Moine Series of Morar, Inverness-shire. Geol Surv. GB, bull. no. 2, 26.
Riedel, W. 1929. Zur Mechanik geologischer Brucherscheinungen. *Centralbl. f. Min. Geol. u. Pal.* **1929 B**, 354.
Ruhland, M. 1958. Allure des plis et plis à axes subverticaux dans les terrains primaires des Vosges méridionales. *Bull. Serv. Carte Géol. Als. Lorr.* **11**, 45.
Rutter, E. H. 1974. The influence of temperature, strain rate and interstitial water in the experimental deformation of calcite rocks. *Tectonophysics* **22**, 311–34.

Sander, B. 1930. *Gefügekunde der Gesteine.* Wein: Julius Springer.
Sander, B. 1948 and 1950. *Einführung in die Gefügekunde der geologischen Körper,* I and II. Wien and Innsbruck: Springer Verlag.
Sander, B. and O. Schmidegg 1926. Zur petrographisch-tektonischen Analyse, III. *Jahrb. Geol. Bundesanst. Wein* **76**, 323, 328.
Sanderson, D. H. 1974. Patterns of boudinage and apparent stretching lineations developed in folded rocks. *J. Geol.* **82**, 651–61.
Scheidegger, A. E. 1958. *Principles of geodynamics,* Ch. 3. Berlin: Springer Verlag.
Scheumann, K. H. 1956. Boudinagen und Mikroboudinagen in Metagabboischen Plagioklas-Amphibolit von Rosswein. *Abh. d. Sachs. Akad. d. Wissensch. zu Leipzig. Math-naturwiss. Kl.* **45**, 1.
Schmidegg, O. 1936. Steilachsige Tektonik und Schlingenbau auf der Südseite der Tiroler Zentralalpen (Austria). *Jahrb. Geol. Bundesanst. Wein* **86**, 115.
Schmidt, W. 1918. Bewegungspuran in Porphyroblasten krystalliner Schiefer. *Sitzber. Akad. Wiss. Wien, Math.-Nat. Kl.* **1**, 127, 293.
Schmidt, W. 1932. *Tektonik und Verformungslehre.* Berlin: Gebrüder Borntraeger.
Scrope, G. P. 1825. *Considerations on volcanoes.* London: W. Phillips.
Sedgwick, A. 1835. Remarks on the structures of large mineral masses . . . etc. *Trans Geol Soc. Lond.* **2** (3), 461.
Seibold, E. 1953. Fiederspalten und Drucksuturen. *Neues Jahrb. Geol. Paläont. Abh.* **96**, 357.
Shackleton, R. M. 1953. The structural evolution of North Wales. *Liv. and Manch. Geol J.* **1**, 261.
Shackleton, R. M. 1957. Downward facing structures of the Highland Border. *Q. J. Geol Soc. Lond.* **113**, 361.

Shainin, V. E. 1950. Conjugate sets of *en échelon* tension fractures in the Athens Limestone at Riverton, Virginia. *Bull. Geol Soc. Am.* **61**, 509.
Sharpe, D. 1847. On slaty cleavage. *Q. J. Geol Soc. Lond.* **3**, 74.
Sharpe, D. 1849. On slaty cleavage (second communication). *Q. J. Geol Soc. Lond.* **5**, 111.
Sheldon, P. G. 1928. Note on the angle of fracture cleavage. *J. Geol.* **36**, 171, Fig. 1.
Sherrill, R. E. 1929. Origin of the *en échelon* faults in North Central Oklahoma. *Bull. Am. Assn Petrol. Geol.* **13**, 31.
Shrock, R. R. 1948. *Sequence in layered rocks.* New York: McGraw-Hill.
Siddans, A. W. B. 1972. Slaty cleavage: a review of research since 1815. *Earth Sci. Rev.* **8**, 205–32.
Sitter, L. U. de 1956. *Structural geology.* New York and London: McGraw-Hill.
Sitter, L. U. de 1958. Boudins and parasitic folds in relation to cleavage and folding. *Geol. en Mijnb. (N. W. ser.)* 20 Jg, 277, 280.
Skempton, A. W. 1966. Some observations on tectonic shear zones. *Proc. 1st Cong. Int. Soc. Rock Mech., Lisbon* **1**, 329–35.
Slater, G. 1927. Studies in the drift deposits of the south-western part of Suffolk. I: The structure of the disturbed deposits in the lower part of the Gipping Valley near Ipswich. *Proc. Geol Assn Lond.* **38**, 157, 176–8.
Sorby, H. C. 1853. On the origin of slaty cleavage. *Edinb. New. Phil J.* **55**, 137.
Sorby, H. C. 1855. On slaty cleavage as exhibited in the Devonian limestones of Devonshire. *Phil Mag.* **4** (11), 20.
Sorby, H. C. 1856. On the theory of the origin of slaty cleavage. *Phil Mag.* **4** (12), 127.
Sorby, H. C. 1857. *On some facts connected with slaty cleavage.* Rep. Brit. Assn Adv. Sci., 92.
Stockwell, C. H. 1950. The use of plunge in the construction of cross-sections of folds. *Proc. Geol Assn Canada* **3**, 97.
Strand, T. 1945. Structural petrology of the Bygdin conglomerate. *Norsk. Geol. Tidssk.* **24** (1944), 14.
Stromgard, K. E. 1973. Stress distribution during the formation of boudinage and pressure shadows. *Tectonophysics* **16**, 215–48.
Sutton, J. 1960. Some structural problems in the Scottish Highlands. *Int. Geol Cong. 21st, Norden, C. R.* part 18, 371.
Sutton, J. and J. Watson 1951. The Pre-Torridonian metamorphic history of the Loch Torridon and Scourie areas in the North-West Highlands, and its bearing on the chronological classification of the Lewisian. *Q. J. Geol Soc. Lond.* **106** (1950), 241.
Sutton, J. and J. Watson 1955. The structure and stratigraphical succession of the Moines of Fannich Forest and Strath Bran, Ross-shire. *Q. J. Geol Soc. Lond.* **110** (1954), 21.
Swanson, C. O. 1941. Flow cleavage in folded beds. *Bull. Geol Soc. Am.* **52**, 1245.

Tanton, T. 1930. Determination of age-relations in folded rocks. *Geol Mag.* **67**, 73.

Tchalenko, J. S. and N. N. Ambraseys 1970. Structural analysis of the Dasht-e Bayez (Iran) earthquake fractures. *Bull. Geol Soc. Am.* **81**, 41–60.

Thurmann, J. 1853. Résumé des lois orographiques générales du système des Monts-Jura. *Bull. Géol. Soc. France* **2** (11) (1853–4), 41.

Treagus, J. E. 1974. A structural cross section of the Moine and Dalradian rocks of the Kinlochleven area, Scotland. *J. Geol Soc.* **130**, 525–44.

Tullis, J., G. L. Shelton and R. A. Yund 1979. Pressure dependence of rock strength: implications for hydraulic weakening. *Bull. Mineral.* **102**, 110–14.

Turner, F. J. 1953. Interpretation of marble fabrics in the light of recent experimental deformation. *Int. Geol Cong. 19th, Algeria, C. R.* sec. 3, pt 3, 95, 109.

Turner, F. J. 1957. Lineation, symmetry, and internal movement in monoclinic tectonite fabrics. *Bull. Geol Soc. Am.* **68**, 1.

Turner, F. J. and J. Verhoogen 1951. *Igneous and metamorphic petrology.* New York: McGraw-Hill.

Turner, F. J. and L. E. Weiss 1963. *Structural analysis of metamorphic tectonites.* New York: McGraw-Hill.

Tyndall, J. 1856. Comparative view of the cleavage of crystals and slate rocks. *Phil Mag.* **4** (12), 35.

Vogt, T. 1930. On the chronological order of deposition of Highland Schists. *Geol Mag.* **67**, 68.

Voll, G. 1960. New work on petrofabrics. *Liv. and Manch. Geol J.* **2**, 503.

Waard, D. de 1955. The inventory of minor structures in a simple fold. *Geol. en Mijnb. (N. W. ser),* 17th Jg., 1.

Wegmann, E. 1929. Beispiele tektonischer Analysen des Grundgebirges in Finnland. *Bull. Comm. Géol. Finlande,* no. 87, 98.

Wegmann, E. 1932. Note sur le boudinage. *Bull. Soc. Geol. France* **5** (2), 477.

Wegmann, E. 1938. Geological investigations in southern Greenland. *Medd. Gronland* **113**, 52.

Wegmann, E. and E. H. Kranck 1931. Beiträge zur Kenntniss der Svecofenniden in Finnland. *Bull. Comm. Géol. Finlande,* no. 89, 62, 92–3.

Wegmann, E. and J. P. Schaer 1957. Lunules tectoniques et traces de mouvements dans les plis du Jura. *Eclog. Geol. Helv.* **50**, 491.

Weiss, L. E. 1954. A study in tectonic style: structural investigation of a marble–quartzite complex in southern California. *Univ. Calif. Pub. in Geol Sci.* **30**, 1, 39 et seq.

Weiss, L. E. 1959. Geometry of superposed folding. *Bull. Geol Soc. Am.* **70**, 91.

Weiss, L. E. and D. B. McIntyre 1957. Structural geometry of Dalradian rocks at Loch Leven, Scottish Highlands. *J. Geol.* **65**, 575.

Weiss, L. E. and D. B. McIntyre 1959. Structural geometry of Dalradian rocks at Loch Leven, Scottish Highlands: a reply. *J. Geol.* **67**, 247.

White, W. S. 1949. Cleavage in east-central Vermont. *Trans Am. Geophys. U.* **30**, 587.

White, W. S. and R. H. Jahns 1950. The structure of central and east-central Vermont. *J. Geol.* **58**, 179.

Whitten, E. H. T. 1966. *Structural geology of folded rocks*. Chicago, Ill.: Rand McNally.
Willis, B. and R. Willis 1923. *Geologic structures*. New York: McGraw-Hill.
Wilson, G. 1946. The relationship of slaty cleavage and kindred structures to tectonics. *Proc. Geol Assn Lond.* **57**, 263, 265–6.
Wilson, G. 1951. The tectonics of the Tintagel area, north Cornwall. *Q. J. Geol Soc. Lond.* **106**, 393, 409, 424, Figs 7, 13a, 13b.
Wilson, G. 1952. A quartz vein system in the Moine Series near Melness, A'Mhoine, north Sutherland, and its tectonic significance. *Geol Mag.* **89**, 141.
Wilson, G. 1953. Mullion and rodding structures in the Moine Series of Scotland. *Proc. Geol Assn Lond.* **64**, 118.
Wilson, G. 1960. The tectonics of the 'Great Ice Chasm' Filchner Ice Shelf, Antarctica. *Proc. Geol Assn Lond.* **71**, 130, Pl. 6.
Wilson, G. 1970. Wrench movements in the Aristarchus region of the Moon. *Proc. Geol Assn.* **81** (3), 595–608, Fig. 3.
Wynne-Edwards, H. R. 1957. The structure of the Westport concordant pluton in the Granville, Ontario. *J. Geol.* **65**, 639.

Zwart, H. J. 1954. La geologie du massif du St-Barthelemy (Pyrenees, France). *Leidse Geol. Med.* **18** (1953), 1.

Index

Bold numbers refer to pages on which terms are defined.

analysis
 geometric 12, 144
 kinematic 13
anticline 19, 20, 21, 22
 micro 59
antiform 22
axes
 kinematic 13
 structural 12, 14
 symmetry 12

boudinage 54, 67, **72**, 82, 104, 108
 of garnets 79
 relationship to principal stresses 76
 relationship to folds 77
 relationship to thrusts 77
 rhomboidal 79
 rotation of 78, 79, 104-5
 tablet of chocolate 73, 78
breccia zone 31
brittle rocks 24, 83

cleavage 35, 41
 axial-plane 36, 42, 50, 59, 65, 70
 de charriage 57
 close joint 37
 crenulation 38, 48
 curved 49, 53
 dislocation 54
 distortion of 95, 96
 false 37, 48
 fan 43, 50
 flow 37, 40, 51, **61**, 70
 fracture 31, 32, 33, 37, 40, 41, 42, 48, 50, 51, 59, 104, 111
 front of 41, 42, 66
 grain 62
 non-pervasive (syn. non-penetrative) 37, 48, 53, 57
 oblique 53
 original 36, 37
 pencil 50, 63
 pervasive 40
 pseudo 39
 refraction 41, 49, 50
 relationship to stratification (bedding) 43, 53, 67
 Rünzelclivage 39
 shear 32, 36, 65
 slaty 12, 40, 42, 51, 61

slice 50
slip 38, 48
 strain-slip 36, 38, 41, 48, 57, 59, 65, 67
 thrust 57
 true 40
corduroy structure 86

décollement, zone of 107
deformation
 ellipsoid of 3
 incremental 3
 multiple 70, 87, **94**, 99
 single phase 70
deformed
 agglomerate 64, 108
 amygdules 64
 fossils 64, 79, 83, 108
 lava pillows 64, 108
 oolites 61, 64
 pebbles 64, 83, 93, 95, 104, 107
 volcanic bombs 108
domes and basins 99
dykes 25, 77, 79

edgewise conglomerate 68
epizone 67

fabrics
 axial 12
 monoclinic 12
 orthorhombic 12
 triclinic 12
fault 27, 41
 faulting, zone of 4
 normal 50, 80
 plane 53
 sheeted zone 53
 tear or wrench 25, 30
 zone 48, 53
feuillets 50
fissility 35, 37
flow structure (igneous rock) 37
fold
 asymmetrical 14, 83
 autochthonous 80, 83
 axial plane 14, 20, 42, 50
 axis 14, 70, 84
 browfold 21
 chevron 59
 concentric 65

Index

congruous 82
conical 15, 70, 84
conjugate 110
crenulation 81
crest 44
cross-folding 99
cylindrical 12, 14, 15, 70, 83, 90
dependent 82
detached 98
diminutive 66
disharmonic 80, 85
drag 68, **80**, 82, 83
elliptical 84
en échelon 85
eye fold 79
flexural slip 10, 65, 80, 82, 102
folding 4
hinge 42, 81
hinge-line 14
incongruous 82
inconstant 84, 91
independent 82
isoclinal 98
major 80, 81
micro 59, 66, 67
minor 80
monoclinic 10
overturned 43
parallel 10, 65
parasitic 10, 67, 68, **80**, 81, 82, 83, 84, 98
plunge 14, 22, 45, 82, 83, 84, 105
pod 84
recumbent 21, 44, 77, 99, 107
régard 14, 43, 80, 83, 95, 104, 108
shear fold 15, 47, 68, 81, 82, 83, 102, 111
similar 65, 68
sinusoidal 59
superposed 99
vergence 14
zig-zag 27
foliation 35, 38

Gleitbretterfaltung 47, 53
gneiss 36, 37
granite gneiss dome 110
Griffelschiefer 50

hogback 34
homogeneous flattening 68
hydraulic fracture 6

igneous intrusion 27, 42, 89, 110

joint
close 48

cross 25, 89
drag 34
feather 27

kink band 33
normal 33, 65
reverse 33, 34

lineation 35, 61, 62, 63, **67**, 70, 86, 89, 95, 98, 99, 100, 109
folding of **99**, 100, 101, 102
lit-par-lit 36
lunules 19

metamorphic aureole 42
metamorphism 42, 62
microlithon 37, 38, 48, 57
migmatization 40, 42
movement picture 12, 62, 107
mullion 68, **86**, 87, 88, 89, 90, 107, 109, 110
fold (syn. bedding) 89
cleavage **89**
irregular **89**, 90
relationship to folds 90
mylonite 109, 110

nappe 21, 83
Alpine 42
Axen 21
Diablerets-Gellihorn 52
gravity glide structure 109
Helvetic 57
Morcles 44, 52

phyllite 62
pore fluids 7, 26
pressure 6
confining 6, 7
containing 24
privileged paths 34, 40
Pumpelly's rule 82

recrystallisation 59, 61, 65
Riedel shear 28
rods 68, **86**, 91, 92, 93, 109
relationship to folds 93, 98
Rünzelclivage 48

salt plug 110
Scherfaltung 47
schistosity 35, 36, 37, 41, 43, 59, **61**, 62, 70
sedimentary structures 17, 45
current bedding 17, 55, 104

graded bedding 17, 44
ripple marks 17
slump structures 55
sole marks 17
shear
  failure 24
  planes 24, 25, 30
  pure 3
  simple 3, 6, 12
  zones 24, 26, 27, 53
slickensides 19, 86, 87
slip, along bedding planes 4, 18–21, 31, 65, 78, 80
spiral porphyroblast 70
standard state 3
strain 3
  ellipsoid 3, 4, 6, 12, 13, 51, 63
  rate 7
stratigraphical succession 17, 18, 31, 59–61, 81, 83, 85, 112
stress 3
  ellipsoid 3, 4, 6, 12, 13
  differential 3, 6, 7, 24, 26, 76
  hydrostatic 3, 26
  non-hydrostatic 3
  principal 3, 8, 25, 77
  shearing 42
structural geology 111
structures (small scale – minor)
  definition vii, ix, 1
  downward facing 44, 45
  superposed 47, 71, 94
  upward facing 63
symmetry
  axial 10, 12
  class 12
  monoclinic 6, 8, 10, 78
  orthorhombic 9, 10, 77, 78
  structural 10, 12
  triclinic 9
  triclinic structure 10, 47
syncline 19, 20, 21
  inverted 21
  Loch Alsh 55
  micro 59
  recumbent 30, 54
synform 22

tectonics vii
  associated minor structures 102
  gliding 83
  inclusion 72
  large scale 102
tensile failure 24
tension gash 26, 27, 31, 77
  conjugate set 29, 30
  en échelon 6, 10, 26, 27, 28, 29, 31
  sigmoidal 6, 30
tension fracture 24, 25
thrust
  boudinage association 77
  cleavage association 51, 52, 53, 54
  fold association 83
  imbricate 57, 83
  minor 83
  Moine 30, 55, 83, 109
  tectonics 80
  thrusting 4, 25, 83
  Sgurr Beag 57
  zone of thrusting 41, 48, 77, 85, 108, 110

veins 25
  tensional 30

work-hardening 65